Tourism and Intercultural Exchange

TOURISM AND CULTURAL CHANGE
Series Editor: Professor Mike Robinson, *Centre for Tourism and Cultural Change, Sheffield Hallam University, UK*

Understanding tourism's relationships with culture(s) and vice versa, is of ever-increasing significance in a globalising world. This series will critically examine the dynamic inter-relationships between tourism and culture(s). Theoretical explorations, research-informed analyses, and detailed historical reviews from a variety of disciplinary perspectives are invited to consider such relationships.

TOURISM AND CULTURAL CHANGE 4
Series Editor: Mike Robinson
Centre for Tourism and Cultural Change, Sheffield Hallam University, UK

Tourism and Intercultural Exchange
Why Tourism Matters

Gavin Jack and Alison Phipps

CBS

CBS PUBLISHERS & DISTRIBUTORS
NEW DELHI • BANGALORE • PUNE (INDIA)

For Gavin's grandparents,
And Alison's parents,
With love

CBS Publishers   ISBN : 978-81-239-1714-6
Channel View   ISBN : 978-1-84541-017-9

First Indian Reprint : 2009

Published by:
S.K. Jain and produced by V.K. Jain for CBS Publishers & Distributors,
4819/XI, 24 Ansari Road, Daryaganj, New Delhi - 110 002, India
e-mail: cbspubs@vsnl.com, cbspubs@airtelmail.in
Website: www.cbspd.com

Branches:
• *Bangalore:* 2975, 17th Cross, K.R. Road,
  Bansankari 2nd Stage, Bangalore - 560 070
  Fax: 080-26771680 • e-mail: cbsbng@dataone.in

• *Pune:* Shaan Brahmha Complex, Basement, Appa Balwant Chowk,
  Budhwar Peth, Next to Ratan Talkies, Pune - 411 002
  Fax: 020-24464059 • e-mail: pune@cbspd.com

Printed at :
Chaman Enterprises, Delhi - 110 095

Epigraph

Ich muß Sie bitten, mit mir in die Unordnung aufgebrochener Kisten, in die von Holzstaub erfühllte Luft, auf den von zerrissenen Papieren bedeckten Boden, unter die Stapel eben nach zweijähriger Dunkelheit wieder ans Tageslicht beförderter Bände sich zu verstetzen. . ...

Walter Benjamin (1973: 169) Ich packe meine Bibliothek aus.

Stories [. . .] are living things; and their real life begins when they start to live in you. [. . .] Stories are subversive because they always come from the other side, and we can never inhabit all sides at once. If we are here, story speaks for there; [. . .] Their democracy is frightening; their ultimate non-allegiance is sobering.

Ben Okri (1997: 44) A Way of Being Free.

Contents

Acknowledgements

We would like to start by acknowledging the support of the University of Stirling's Faculty of Management which funded most of the fieldwork on which this book is based, as well as the University of Glasgow which provided money for recording equipment. Time for the writing of this book was made materially possible by the AHRB and the University of Glasgow (for Alison), and the University of Leicester Management Centre (for Gavin).

For all those tourists, and locals, that made our fieldwork on Skye a chocolate box of delights, we are eternally grateful. Some of you know who you are; some of you will never know. We hope you enjoy the stories recounted in this book and that you take pleasure in recognising yourselves in them.

The wonderful team at MLM and Channel View Publishing, especially Mike Grover, Marjukka Grover, Sami Grover and Sarah Williams, have been a constant source of support, inspiration and patience. It is a pleasure to work with you.

We would particularly like to thank our editor Mike Robinson. Our work has benefited tremendously from his kind and considerable intellect, and his ability to give incisive and valuable criticism.

Valérie Fournier read the draft manuscript in its entirety. Her generosity in doing so is very much appreciated, and she has contributed substantially to the development of our thinking.

We were fortunate to be able to present earlier versions of this work in the Centre for Language and Communication Research at the University of Cardiff. We would like to thank Crispin Thurlow and Adam Jaworski for their kind invitation and special hospitality. We also acknowledge the help of Ron Barnett, Fiona Anderson-Gough and Karen Dale in clarifying our writing. Sharon Macdonald provided invaluable advice in the early stages of this work.

Outwith academic life, our partners, Ian Kearns and Robert Swinfen, have demonstrated considerable patience, abiding faith and even some enthusiasm for our project.

This book is dedicated to the people who inspired it in the first place. Thanks, gran and grandad, for the Highland hospitality. Thanks, mum and dad, for a great idea. And finally, thanks to Felix (*Fratercula arctica*), tourist *par excellence*.

All omissions and errors remain our responsibility.

Chapter 1
Why Tourism Matters

Tourism Matters

Tourism matters. It matters not because of the dystopian voices that call it a blight on the planet, or the desires of the functional servants of capital and/or the state that rush to tourism as economic saviour. Tourism is much more important than this. Tourism matters because it provides both a lens onto and an energy for relationships with everyday life. It invites us to engage in exchanges of life with others, and to remind us thereby of its most precious and vulnerable aspect: the intricate relativities of defining people who are not us. Tourism matters because, in a world of confusing connections and disconnections between human beings, our lives with others matter.

Exchange and Authenticity

People tell big stories about tourism. In Dean MacCannell's (1976) *The Tourist* for instance, we are presented with something of a grand narrative of tourism, that is, an overarching story about tourism as a search for authenticity against the background of the alienating conditions of working Modernity. Tourists play the role of 'alienated moderns' ensconced in the pursuit of 'authentic', 'real', 'whole' social relations denied by their position as labour in capitalist relations of production.

It is easy to see how this early work of MacCannell's might be considered a story of differentiated Modernity (Meethan, 2001). It paints a picture of the labourer toiling in the workhouses of industrial capitalism and then seeking retreat in changing forms of leisure. This includes 1930s and 1940s seaside resorts, 1950s holiday camps, 1960s and 1970s Mediterranean package holidays and now, in the contemporary diversity of travel possibilities, perhaps cheap flights for short city breaks, or tailor-made adventure holidays. Whatever the escape, the work–leisure binary, itself a modern idea, compared, say to premodern notions of work and rest in dialectical tension, shapes many a tale about the position of the tourist subject in Modernity.

In addition to the work–leisure binary, a further two elements are key to this story. First, there is the notion of capitalist relations of exchange

1

and its constituent relations of production (the labour process) and consumption; and second, the trope of the Other and its authentic social relations. Both these elements suggest that tourism is a quest for the Other born out of the political economy of capitalist exchange relations and its differentiation of work and leisure, the public and the private spheres.

However, these two key elements, as presented in this early work of MacCannell, are problematic. Indeed, criticism of this early work of tourism is now well rehearsed. Toeing a neo-Marxian line, MacCannell's implicitly reductionist, abstracted and deterministic view of the tourist was always likely to be fundamentally undermined in subsequent sociological and anthropological arenas as these turned to questions of language, culture and representation inspired by poststructuralist and postmodern social and cultural theory. Here the tourist as the passive dupe of the labour process seems conceptually problematic.

A further key concern in tourism's literature is the concept of authenticity. Drawing upon Graburn and Bruner, Edensor (1998) points out that MacCannell's analysis serves to reify the concept of authenticity, in the process disconnecting any idea of it from the social, cultural and historical contexts from which it might emerge. Rather than some kind of essential or universal category of analysis, Edensor argues that we might be better served through a dynamic and emergent conception of authenticity that is attuned to the subject positions and historical and geographical settings of those talking about it. So, for example, the way in which indigenous peoples are now making crafts for tourist markets rather than for their own domestic needs might cause us to re-think our notions of the authentic.

To examine authenticity in terms of such subject positions is problematic because it conflates ideas of authenticity with alterity. In other words, it assumes, as in the above example, that alterity – Otherness – can be easily categorised as either modern or nonmodern. And yet, as our example demonstrates, the 'modern' entrepreneur here is precisely the one who is being consumed by the 'modern' tourist as somehow authentic and indigenous.

MacCannell's rather narrow view of authenticity, and for that matter, the fate of the labouring subject, is an outcome of the structures of the grand narrative he presents of modernity, that most particular of Western experiences of the development of industrial capitalism. In terms of narrative structure, MacCannell's analysis is possible because of its deployment of a dualistic mode of thinking which holds in place a set of binary divisions. Such thinking often results in the pigeonholing of social phenomena, tourism included, into 'either–or' categories e.g. tourism is bad, unethical and subjugating rather than good, ethical and emancipatory. Given that both MacCannell's work, and this dualistic mode of analysis would seem to be indicative of tourism research in general, it is hardly surprising that the kinds of positions in the tourism literature collected together in Box 1 have gained privileged discursive currency. For instance, journals of ethical tourism typically presume the inherent 'goodness' of their positions rather than examining the cultural dimensions that create a discourse on ethics in tourism in the first place.

This far from exhaustive list represents tourism, in the left-hand column, as an undesirable outcome of the socioeconomic basis of Modernity – a product of commodified relations between producers and consumers, an environmentally and ethically dubious set of social relations, an emerging global phenomenon, an alienating experience for the passive dupes of Modernity, a destroyer of culture(s). If this is true, then tourism can only ever be evil. And if the positions in the right-hand column of Box 1 are true, then this picture suddenly becomes the converse, utopian even. These frameworks and assumptions drive certain lines of analysis and questioning, leading to the characteristic stories we tell of tourism for good or ill. Breaking with such modes of thinking, however, thinking which Lévi-Strauss has taught us to

Box 1 Positions in the tourism literature

Tourism is either cast as ...	Or ...
Commercial bubble	Economic saviour
Commodification	Sustainable
Unsustainable	Ethical
Colonial	Ecological
Detrimental	Restorative
Homogenising	Developmental
Fake	Educational
Imperialist	Connecting
Unethical	Agentic
Separating	Oneness
Unecological	Intercultural nirvana
Greedy	
Disconnected	
Alienating	
Creating dupes	

understand as fundamental to the logic of Western thought and culture, is not at all easy.

These lists can be used to construct exactly the kinds of sweeping pictures of the works that constitute a literature on tourism of which one might be critical, in the same way as we have been critical of MacCannell's research. Our intention, though, is largely a rhetorical, evocative positioning of this work within the field of tourism. There are of course a number of diverse studies and approaches that critique and offer alternative visions of tourism to those presented in Box 1. We return to these in more detail in Chapter 2.

Postmodern analyses of tourism attempt to pick apart this knotting together of tourism, commodified exchange relations and the search for authenticity. They show the playfulness of the tourist subject, resisting the structures of global capitalism and reading ironically the privileged sites of tourist destinations and other so-called stages of tourism (Kirschenblatt-Gimblett, 1998). Similarly, several recent pieces reflect critically on the stranglehold of quantitative work on the production of tourism research, and call for a more theoretically sophisticated approach to its study (Franklin & Crang, 2001; Koshar, 2000; Meethan, 2001).

Despite these conceptual questionings, the story of a differentiated Modernity, that most periodising and spatialising of stories, has created a lens through which many sociological analyses have been refracted. And if anything, the deployment of postmodernism serves to offer us yet another grand narrative dressed up as new social theory in the area of tourism. It is a scepticism, which itself becomes a grand narrative.

In this book, we align ourselves with calls not only for more theoretically driven approaches to tourism research, but with research that calls into question some of the key assumptions and tropes of the big story of tourism as a consequence of the differentiation inherent in Modernity (Meethan, 2001). In this latter regard, we follow Koshar (2000), who suggests that tourism studies, far from requiring grand narratives, will be better served by the development and grounding of concepts. As Foucault (1978) reminds us, discursive technologies can have the duplicitous effect of rendering things both visible and invisible – we can therefore only see certain things through the lens of specific frameworks, such as those we visited in Box 1. However, this is as much a choice as an inevitability. We can and have to read old stories in fresh ways, as Bauman (1998: 5) reminds us:

Not asking certain questions is pregnant with more dangers than failing to answer the questions already on the official agenda; while asking the wrong kind of questions all too often helps avert eyes from the truly important issues. The price of silence is paid in the hard currency of human suffering.

What would happen, for instance, if, returning to our earlier box, we take neither the position that tourism is only evil nor the opposite view of an intercultural, ethical or economic nirvana? What if we render suspicious some of the binaries that frame much analysis in tourism? What can tourism become if we ask different questions? What if we try working critically both with, and against the grain of universal stories? What might happen for instance if:

- We look at tourism not as a reification of a set of social relations, but as an outcome of social and material processes and practices in everyday life?
- We understand tourism not just as an exotic search for the Other, but also as a banal activity that enacts the routines of everyday life across the binaries of work and leisure?
- We see tourists living through different experiences, not as products of either micro- or macrostructures?
- We consider the influence of the traditions in which we come to work on tourism and with tourism?
- We examine the emotion, the work of the imagination and the sheer buzz that comes from and generates tourism?
- We consider the way tourism is animated by narrative?

In short, how else might tourism, understood as a participatory set of interactions-in-the-world (as opposed to a search for, say, authenticity) look beyond the narrative confines of the stories already told by others about the dualistic micro- and macrostructures and behaviours of economic and cultural relations in late capitalism?

About this Book

This book presents work that addresses these concerns. Through it we attempt to engage the traditions of tourism research and the understandings of tourism handed down to us. We endeavour to ask how the material and affective dimensions of tourist experience are played out in the myriad interactions of everyday tourist life. Crucially, the book sets out to address these questions by exploring a neglected area of research in tourism studies, namely intercultural communication and the way this is constructed and mediated through material culture and through language.

The fact that tourism is an intercultural activity, constructed within and through language, has been largely ignored in tourism research until very recently. This is unfortunate, since a study of tourism as a form of intercultural communication could be instructive in relation to the notion that tourism involves, to repeat, a quest for a participatory set of interactions-in-the-world, *and not, importantly, for authenticity*. Authenticity has become a perennial cul-de-sac in tourist research, we would argue.

The term 'intercultural communication' usually encompasses the notion of interaction between members of different cultures, understood loosely as different national groupings. We use the term in this sense throughout this book but we would want to add certain nuances to this overarching definition. Because we understand intercultural communication to be a participatory set of actions in the world, we see it as taking form in dialogical and material exchanges between members of cultural groupings. We see cultural membership as marked variously by race, ethnicity, nationality, language, class, age and gender.

We neither subscribe to utopian nor dystopian views of intercultural communication. Intercultural communication can take many forms from extreme conflict to peaceful harmony. This is as true for international politics as it is for a brief encounter between two tourists in a youth hostel kitchen. The intercultural communication literature is studded with examples of attempts to model positive and negative intercultural interaction and to suggest ways of ameliorating encounter. This literature has its place but is not our concern here. Instead we treat intercultural communication by and large phenomenologically. That is to say we consider intercultural phenomena as they present themselves to us within the modes of tourism.

Tourism provides a particularly concentrated and significant occasion for intercultural communication, and for a potential mixing of different social groupings. In this mix we find relations that are inflected by power relations in highly nuanced ways, as resistance, as domination and as subtle shifts between people as they meet and exchange. Conceptualising tourism as a site and form of intercultural communication, this book addresses these broader perspectives through the following questions:

- What facilitates/hinders intercultural communication for tourists?
- To what extent may the concept of 'exchange' and its particular structure be seen as a key to understanding intercultural communication in tourism?
- What is the relationship between material life, tourism and exchange?
- What is the relationship between material life and the emotional experiences of tourism relayed in stories, big and small?
- What lessons may we learn through tourism about nontourism?

In order to address these questions we present an account of an ethnographic study of tourism on the Isle of Skye, an island situated just off the northwest coast of Scotland. As part of Scotland's Highlands and Islands, the Isle of Skye is one of Scotland's top destinations for overseas tourism. Each year it attracts a large number of several key national tourist groupings, most notably Americans, Germans, French, Italians, Dutch and Canadians. As such, the island offers plenty of potential opportunities for witnessing and engaging in intercultural communication. It should be noted that the Isle of Skye has been the site of several other ethnographic and geographic studies addressing different aspects of life on the island (Brougham & Butler, 1975; Macdonald, 1997).

The ethnography was not a protracted affair. Rather than hanging about the island for a whole tourist season collecting data, we decided to act as 'real' tourists in 'real' tourist time. In other words we planned a holiday and spent the same amount of time on the island as regular tourists. We booked a number of different kinds of accommodation in different parts of the island and let ourselves be guided to particular attractions and tourist spots by the unfolding interactions of the holiday. Chapter 3 of this first Part deals in greater detail with questions of method and methodology. We might perhaps emphasise here only one or two key details about ethnographic method.

First of all, our praxis as 'ethnographers' was an engaged and active part of the very exchange-based processes that it sought to study. We were very much participants as well as observers in that which we studied. We were both tourist ethnographers and ethnographer tourists. Indeed the similarities between the material lives of ethnographer and tourist led us to understand aspects of tourism as a material, representational, and as an ethnographic practice.

Secondly, the ethnography was conducted jointly and, in many respects, might well be described as a form of joint travel writing. We travelled together ostensibly to others as a couple, and visited the same sites, the same accommodation. We kept separate ethnographic journals and collected separate artefacts and documents. However, working together gave us the opportunity to have discussions and tell stories of what we observed and the events in which we participated. We tape-recorded most of these joint conversations and these became important resources for the work.

The approach is, above all, *dialogical*. This dialogical nature, which forms part of the essential structure of exchange and tourism, has also formed the basis of our working method, which we might term 'exchange ethnography'. Tourism is a shared, social activity and we have sought to mirror this in our joint ethnographic work. It is hoped that the style of the book may reflect this dialogical aspect by representing experience in a variety of ways stylistically.

The Travel Bag

The book works, structurally and thematically, with the analogy of the travel bag and its packing, unpacking and repacking. Although used as a metaphor to aid the troping of our discussions, the travel bag also has a material reality and we will relate the analogies to material objects and cultural 'baggage'. The material and the analogies are examined for their role in creating instances of speech acts in the form of other language use and intercultural communication.

The 'baggages' that are brought into tourism thus form a focus of the analysis. The importance of baggage is discussed in the context of tourism as a process of identity formation and of everyday interactions between people, places, stories and things. This enables a continuing return to our question regarding what facilitates and hinders inter-cultural exchange in our particular tourist contexts. Each of the three key substantive sections of the book addresses a different part of the inter-related processes of dealing with baggage, that is to say in the context of this book: *packing, unpacking and repacking.*

In folding out these conceptual distinctions, we combine ethnographic description with chosen pieces of theoretical insight from a variety of authors in order to give texture to the work. Throughout we analyse and extend the key concept of exchange, seeing its particular structure emerge in dialogue, in economies, in narrative, in performance, in material life, in languages, translation and intercultural communicative situations.

Stories

This study works with grounded concepts of intercultural encounters in tourism, not so much as semiotic exercises, as has been the dominant view in social theories of tourism, but as complex narrative exchanges situated in a web of different signifying practices, spatial contexts, material contingencies, and social and cultural formations. In addressing questions of intercultural communication we engage a wide range of concepts from different disciplinary fields.

It is not our intent to tell the story of the tourist as *a priori* cultural or economic dupe but to discuss alternative practices that come to light through research questions focusing on the interpersonal and intercultural contacts that intercultural communication can facilitate. Nor does this research seek to establish prior clear boundaries between 'hosts' and 'guests', 'tourists' and 'travellers'. Such binary categorisations, to repeat, have their place in the history and theorising of Modernity and its emergence. In the supercomplex (Barnett, 2000), supermodern (Augé, 1995), even 'liquid' (Bauman, 2000) conditions of our present, stories of others may emerge in different, even new forms. Stories move with people, they surround material life and this movement gives rise to changing forms. Form, according to Williams (1977), always has an active material base.

Stories and narratives, then, are a theme throughout this book. Narratives, we found, were crucially enabled by the practices and objects of everyday material life. Material life and its practices, we argue, are inextricably linked to narratives within the tourist context.

Tourism as Intercultural Communication in Scotland

In examining tourism to Scotland we are breaking with the tendency in both the anthropology of tourism, and in tourism studies more generally, of focusing firmly on tourism that occurs along the North–South axis. Instead we are examining tourism that occurs in the West. We are also examining tourism that does not search out the sun. In other words, we are attempting to move against the flow of the main studies of tourism in anthropology.

Anthropologists largely came to the study of tourism as a nostalgic, salvage activity. They began to describe and reflect upon the major cultural and social changes wrought amongst those they were studying, to document the changes and to take up a political stance alongside the 'victims' of change. The very notion of salvage ethnography comes freighted with assumptions of good and bad cultural processes, or good and bad cultures, and a belief that what is to be saved is an unquestionable good. As a result, the anthropology of tourism has brought a particular perspective to the study of tourism that has struggled to break with this nostalgic view of other cultures. It has also struggled to examine the nuances of power that attend such championing of other peoples. Indeed, as Bartlett (2001) points out, it may well be that the champions of cultural salvage are not at all welcomed by those they are attempting to save.

By turning our attention to Scotland, we wish to examine the different and varied intercultural phenomena that occur as a part of the overall picture of Scottish tourism. We wish to do this by trying to take an alternative route to the well trodden North–South path and in so doing we would align ourselves with the recent trends in applied anthropology, auto-anthropology and of anthropology done at home.

Choosing Scotland as our focus enables us to enact several positions. It allows us to practice participant observation by being tourists at home and by being 'Scottish' hosts. It allows us to move in and out of other European languages without leaving the predominantly Anglophone context of home. It enables us to see the other in ourselves, noticeably the other of the Highlands and Islands, the Gaelic-speaking minorities. It enables us to be both one thing and the other.

In addition, our focus on Scotland throws us into a different relationship with other cultures, and other social groupings, to those that formed the basis of salvage ethnographies of the past. It shifts the nature and balance of intercultural power relations and forces us to confront questions about the commodification of our own culture as well as the commodification of the culture of others. In short, it forces us to communicate interculturally, in multiple ways and from multiple positions.

In Scotland, as elsewhere, the history of tourism is one of elite travel. In Scotland, as elsewhere, we find a wide variety of tourism-shaped activities ranging from seaside resorts, to heritage sites to remote scenery. In Scotland there are packages offered and there are opportunities for independent travel. Scottish tourism takes all sorts of people and in this respect it is no doubt possible to generalise about tourism based on a study of tourism to Scotland.

Equally, however, tourism in Scotland shows marked distinctions, some of which it may share with other destinations, but others that it may also claim as unique to the place. The latter are framed metonymically – that is to say, there are certain things which stand for Scottish tourism that could not stand for anywhere else, and these communicate culturally and interculturally: Shortbread, tartan, bagpipes, haggis, a distinctive landscape of lochs and glens, Bonnie Prince Charlie.

Metonymical communication is not equally powerful with all people. Those drawn by the tourist destination that is Scotland are responding to fragments of culture that have come to stand for a whole. In our own study we encountered Italians, Dutch, Danish, Spanish and French tourists numbered among other Europeans, and Canadians, Australians and Americans, but this metonymical power of attraction was most strongly in evidence among the Germans and the English. This is also borne out by annual statistics. To this we should also add that the tourists in these cultural groupings broadly shared social class. Some were students and young families youth hostelling, others were older, all possessed considerable amounts of educational capital, though possession of economic capital varied considerably.

It is worth elaborating here, but with a note of caution. With Ingold (1993), Keesing (1994) and Eagleton (2000) we are sceptical of the benefits of a continued overuse of the term culture or the essentialising and intercultural modelling of so-called cultural behaviours and national stereotypes. The idea of culture, as a concept, has indeed become so overweening and overwritten as to have become a problematic term in analytical contexts. As Ingold (1993: 230) puts it:

> 'culture' has continued to swim with the tide of intellectual fashion, leaving behind it an accumulating trail of discarded significances not unlike a pile of old clothes.

However, the term culture, used metonymically, is also a commonplace and useful shorthand for referring to peoples and nations. The nation state, its origins and constructed nature have been the focus of very considerable debate and deconstruction and we laud those attempts at reaching into the mess of human relation and human identity, difference, construction and projection (Anderson, 1991; Hall & du Gay, 1996).

Much scholarship, particularly in the quantitative social sciences, has tended to essentialise, and to ask questions that are based on *a priori* commonsense categories (Bauman, 2002). However, equally problematic is the opposite tendency to unpick until there is nothing meaningful left to say. Again, Ingold (1993: 217) is instructive:

To sign up to the project of avoiding ethnocentrism is therefore to make an unequivocal assertion of superiority over the run of ordinary humans, patronizingly known as 'cultural members' or 'informants', and to do so in a strikingly western idiom. It is the characteristically anthropological expression of the West's symbolic appropriation of the rest.

Nation states are indeed imagined communities. Ethnographers may indeed return from their cultural fields with little more than 'a haversack of fictions' (Read, 1993) and yet, we would contend that it is these very imaginings and these very fictions which provide links into common experiences of material life.

Again with Ingold (1993, 1994, 2000), we focus not so much on the way in which cultures are constructed, but what the world looks like from certain dwelt-in viewpoints. This perspective enables us not so much to project a prior category of Germanness or Italianness or Americanness onto all those we encounter who speak German or Italian or American English, or carry what we *perceive* to be quintessentially German, Italian or American objects. Rather it enables us to understand these people as living in a reciprocal relationship with materials and discourses that are most commonly, though not only, encountered within the political, economic, geographic and linguistic spaces that we have been educated, through media and schooling and upbringing, to recognise as belonging to these cultures.

The story of how these particular cultural groups – notably the Germans and the English – come to visit Scotland in such numbers can be a complex one. We do not presume here to make a case for determinism, to isolate sets of instances and behaviours which will mean that we may generalise from these instances into all of human life subject to these circumstances. It is always a danger in hermeneutic work, given the dominance of positivistic paradigms, to be suggestive of the idea that 'as people who live here do this, all people who live in similar circumstances will do similar things'. We are acutely aware of the danger with this work that we will be interpreted as saying that 'all Germans love Scotland' and that they do this 'because of *Braveheart* and *Macbeth*' or that 'all Americans come to play golf and search for their ancestors'. This is not our interest.

Let us take the instance of German tourism to Scotland for a moment. All Germans do not love Scotland. Not all Germans want to travel on holiday to Scotland. Not all Germans have fan sites on the subject of Scotland. Not all Germans have seen *Braveheart* or read *Macbeth*. In what we may term, for simplicity's sake, German culture, there is as much a fascination for the South (in the German context, to be understood as the sunny climes around the Mediterranean) as there is a draw towards the Romantic myth of the North of which Scotland is part. This myth takes form in the narrative and structures of the works of Thomas Mann and is evident in Weimar Classicism. However, it also sits alongside the desire for a place in the sun, which exercises the same kind of pull in German culture as it does in many other Western cultures.

The point is perhaps a subtle one, but it is important in the context of opening up what will be our later argument regarding the operation of narrative and the triggers for narrative practice. We contend here that material life provides a constant source of opportunities for intercultural exchange; for narrative, for memory, for reading, sensing, for the emotional work of interpreting and communicating. Taking the intercultural phenomenon of tourism in Scotland as our starting point means, of necessity, that we have to contend with national cultures, but to attempt to do so without denying their specificity and peculiarity, or reaching for essentialist models. The fact that cultural fragments of materials cluster in certain places, at certain times, for long periods of time – geological time even – or as ephemera, in scarcity or abundance, is important in providing markers of potential alterity for tourists.

When we come to examine the contents of our metaphorical travel bags, we are not rifling through stuff to find the quintessential German, Italian or American objects. But we are concerned to see how people, living in certain times, places and social spaces, with their senses of selves and senses of others, have come to take a trip to Scotland in numbers that are statistically significant. We are interested in the practical aspects of dwelling alongside objects, memories, language systems, myths, and the ways these interact to form relations. In short, we are concerned with the material life of exchange.

The material life of exchange is intimately connected with the narratives and stories people live with, and by, and generate themselves. In other words, it is not enough to maintain that human worlds are culturally constructed, far more, it is our contention, drawing on Archer (2000) and Ingold (2000), that *Homo Faber* is engaged in constant acts of making the world out of material encounters, engagements and exchanges. How this key element of exchange underpins our work is the subject of our next chapter.

Chapter 2
The Give and the Take

Although the concept of exchange has traditionally been central to a number of disciplines ranging from anthropology and economics, to social psychology and marketing, it is not for reasons of its disciplinary ubiquity that we have chosen it as key to the conceptual framework discussed in this chapter. Rather, we privilege the concept of exchange because it was a recurring and central practice in the tourist matters we surveyed and participated in. In the remaining sections of this book, we ground the concept of exchange in ethnographic detail. For the purposes of this chapter, however, it is useful to present and debate academic work on exchange in order to clear the ground for later.

Tourism as Exchange: An Introduction

Anthropological research tells us that exchange is a universal human activity (Mauss, 1990). It is found in all societies and communities, whether large-scale market-based societies, or small-scale subsistence societies, and across all times. Our everyday lives are filled with different kinds of exchange; not only commercial exchanges when we swap money for desired goods and services, but also exchanges of love, friendship, community. Exchange can be predictable, the stuff of habit such as the visit to the (super)market, but it can also be unexpected, fun, unusual and uncertain. Exchange does not, however, exist in a social vacuum. Mauss (1990) famously identified ways in which exchange between human beings is both enabled and constrained by a number of societal and cultural forces, organised variously as market structures, social hierarchies or cultural values. As John Davis (1992: 1) puts it:

> Exchange is interesting because it is the chief means by which useful things move from one person to another; because it is an important way in which people create and maintain social hierarchy; because it is a richly symbolic activity, all exchanges have got meaning; and because for Britons and many others it is an important source of metaphors about social relations, about social order, about the fundamental processes of nature.

14

This chapter begins with the premise that exchange is both a key *practice* of tourists and a useful *metaphor* for understanding the nature of tourism as an everyday social activity. In terms of the former, exchange can be viewed as both a structure and an action of the social and symbolic relations that constitute tourism. And in terms of the latter, it provides us with a concept whose dimensions and machinations can be grounded and articulated through the details of empirical study.

Whilst acknowledging exchange as a universal human activity, it might be suggested, however, that it is a social practice that takes divergent cultural and institutional forms which have emerged in historically and societally specific ways, and whose meaning is contextually embedded in certain spatiotemporal relations. This means two things in terms of its theorisation. First, that notions of exchange and its organisation have been subject to the impulses of many theoretical grand narratives told about the emergence of modern society. As Ning Wang (2000: 17) reminds us, 'the emergence of the tourist has to do with the *enabling* conditions of modernity' (sic), a suggestion which has framed the notion of exchange within a very particular set of theoretical (and primarily sociological) vocabularies and traditions of research in a variety of disciplines, which encourage a reading of tourism against the development of capitalist relations of production.

The second notion, that exchange is contextually embedded, impels researchers to investigate exchange not so much through the big lens of the grand narrative, but through a microempirical investigation of exchange as a social and cultural practice. If you like, this casts a small lens onto the processes of exchange within everyday tourist life. These two approaches have tended to predominate the manner in which researchers have come to understand the cultural economy of life (du Gay & Pryke, 2002) across a number of humanities and social science disciplines.

This book pursues the study of tourism not so much as either a macro or a micro phenomenon, whereby the one is set dialectically against the other, rather it attempts to work with and through the variegated life of exchange. We see exchange, then, as sets of social practices culturally located.

In this chapter we begin by arguing that the notion of exchange in tourism studies has too often been confined to the kinds of sociological grand narratives instantiated by the first kind of approach above. We suggest that whilst such big stories certainly facilitate some understanding of tourism as exchange, they are also somewhat restricted in large part due to the kinds of dualistic forms of analysis that characterise these accounts. In the latter regard, we argue that exchange has often too hastily been ascribed meaning within capitalist exchange relations, where it becomes an abstracted and decontextualised social practice that reduces the role of the tourist to the passive assimilator of dominant ideology.

We attempt to move outwith a sociological understanding of tourism exchange to investigate how ideas from anthropology, material culture and cultural economy might be used to inform an alternative understanding. Drawing on this wider range of literature, we create a conceptual framework that sensitises us to a broader understanding of the term exchange, one that is grounded in its material, emotional and cultural nature. We see exchange as fundamental to intercultural communication in the tourist context. We begin though by looking at how exchange might be theorised against the development of modern society.

Exchange and Modernity

There are, as we saw earlier in Chapter 1, different kinds of stories to be told about tourism. Whether we subscribe to MacCannell's early work on tourists as alienated moderns, Urry's visually prioritised tourist system of the personal and social tourist gaze, or Ritzer's explorations of McDonaldisation and McDisneyisation respectively, what we get in all of these is a story about the nature of tourism as a macro social and economic system, and the place of the tourist subject within it. These grand narratives provide overarching structures of intersecting economic, sociospatial, temporal and cultural dimensions through which large-scale theories of tourism are interpreted. Perhaps more tellingly, each of these narratives, especially MacCannell's and Ritzer's, articulate different relationships with neo-Marxist, neo-Weberian and critical theoretical arguments. As such these are grand narratives that make explicitly temporal claims in the sense that they conceptualise tourism with relation to Modernity and the historical development of the Western world in social and economic terms. These are stories that, as we have seen, are largely negative, casting tourism either as a consequence of alienating and rationalising modernism, or as recompense to it, or both. In short they can both be read as dystopian tales of the subject under Modernity.

Whilst these grand narratives are only part of an emerging and relatively heterogeneous tourism literature, which has place for an alternative focus on the cultural practices of tourists and a focus on their experiences and motives, these grand narratives continue to inform much scholarship in the area. Taking up our earlier analysis from Chapter 1, we concentrate on the way exchange, within the context of Modernity's tourism, has been interpreted through two related, but slightly differently accented sociological grand narratives of tourism: that of MacCannell, as commented upon by Meethan, and then Ritzer's accounts of tourism as the extension of the principles of McDonaldisation, which he labels McDisneyisation.

Meethan (2001), with implicit reference to MacCannell's early work, represents very elegantly a sociological story of now received wisdom about tourism as a search for authenticity against the background of the alienating conditions of working Modernity. It goes something like this. A key characteristic of Modernity is differentiation: the creation of public and private spheres, divisions between the world of work and the world of leisure *inter alia*. In the work sphere, Modernity proves a disruptive force to 'real life' – as workers sell their labour power (i.e. their labour time) to the owners of production for a wage, the labour process alienates them from the conditions of their exploitation whilst at the same time thwarting the expression of a collective class consciousness through which this exploitation could be known and overthrown.

Through the labour process, human creativity and 'authentic' relations with others are disrupted. We are alienated from our own labour, and disengaged from those around us. We then use our leisure time to deal with our alienation, to rediscover the authenticity of a life lost in the alienating conditions of workaday life. The site and playground for the alienated modern in search of the 'wholeness' absent from everyday life are nonmodern or premodern societies (by definition undifferentiated and therefore a place of 'real and authentic' social relations).

In this clearly neo-Marxist view of the development of modern society, tourism is intricately linked to exchange framed in terms of capitalist relations of production, and in terms of authenticity bound up in an image of others who belong to nonmodern, nonindustrial and nonalienating societies, available for our temporarily leisured consumption. In such a story, tourism is a commodity relation, promising an authenticating consumption experience that can be purchased in exchange for the money earned through the wage labour process. Tourism, then, is an act of consumption, not of production, and one wholly founded on the search for authentic self–other relations. This view has been influential in the study of tourism. John Glendenning's (1997) work on Scotland points to exactly this theme. He argues that travel writing introduced Scotland to potential tourists as the ground of genuine experiences. To quote him:

It is true that tourism and touristic Scotland have served as romantic displacements of the rigors of modernity and that the self that is asserted thereby can be highly fictitious. (...) In the modern world abides a sense of the loss of true depths and a corresponding restlessness that motivates not only tourism but science, technology, and economics generally. Tourism promises fullness of experience, abundance of life, and sometimes it fulfils its promise. It is most likely to do so when it is pursued with few preconceptions, with the desire to learn, with an eagerness for dialogue, with determination that there and then will not supplant here and now, and with a realization that any moment, even at home, is potentially as deep and mysterious as Loch Katharine. (Glendenning, 1997: 238)

In sociological circles, however, the reduction of tourism to a direct outcome of economic production is problematic. It is implicitly based on a particular reading of Marx's base-superstructure model, or at least his articulations of the relationship between the economic structure of society and its social, political and cultural life (for instance, see Marx, 1859). According to such readings, all political, social, cultural and intellectual life becomes an epiphenomenon of the economic apparatus, reduced to the role of ideology and blinding large segments of society to the conditions of their dependence and exploitation. Not only is this to presuppose culture and the economy as independent spheres of life, it is also to subjugate expressions of the former, to the interests and workings of the latter. Max Weber (1978), in his work *Economy and Society*, is critical of such a view of economy–culture relations and the monocausality of economic structures. He describes the dependencies of culture and economy as part of the notion of the multiple determinations of society. Of interest to Weber is the cultural and religious background to the development of capitalist relations of production, and the manner in which the former and the latter were mutually related (best exemplified in his work on the Protestant Work Ethic). Whilst Marx and Weber's works differ in a number of ways (e.g. by reading Marx's present as a dystopic one, and his future utopic; and Weber as just plain dystopic), both would seem to agree on the corrosive nature of Modernity.

In relation to the study of consumption and tourism as an instance thereof, Weber's work on the rationalisation of society and the concept of the 'iron cage' has provided a useful and much drawn upon framework for analysis. What is interesting about these notions, in comparison to the monocausal readings of the base-superstructure notion, is that they theorise consumption and its attendant social and economic relations, within a *cultural* frame of analysis where ideological interests come into play. In relation to tourism, this gives us another grand narrative wherein the growth of tourism, and its various structures and practices of leisure, is taken as evidence of the problematic rationalisation of a growing number of societies and communities. The use of Weber's ideas on rationalisation, and its relationship to the modern tourist subject, has of course most famously been taken up in the work of George Ritzer.

The Post-tourist

George Ritzer's original McDonaldisation thesis (1996) takes as its point of departure the view that the fast-food chain McDonalds has usurped bureaucratic structure as the model for rationalisation in the modern world. He argues not only that a certain form of rationality associated with ideas of efficiency, calculability, predictability and control is characteristic of the way in which McDonalds conducts business, but also that such rationality is pervading an increasing number of economic sectors both within the USA and in other societies across the globe. He presents a wealth of evidence to support his argument about the spread of the McDonald's logic across sectors as diverse as travel and tourism, health care and education. In addition to documenting this phenomenon, Ritzer is also critical of it. Indeed, his thesis is rather a dystopian one (akin to Weber's) in so far as he suggests that an increasing number of spheres of everyday life are being subjected to and thus disenchanted by the standardising, dehumanising, impersonal and calculable logic of McDonalds. Such a logic is for Ritzer the central 'iron cage' (to use Weber's phrase) of the contemporary world through which the charisma of everyday social life has come to be routinised and rendered predictable. In terms of these consequences, there lies, for Ritzer at least, a profound 'irrationality' at the core of the spread of 'rationality'.

In this original book, tourism is given by Ritzer as an example of an economic sector affected by the spread of the McDonald's logic. In this regard, he uses package tours, amongst other things, as evidence of the McDonaldisation of tourism where tourists whiz around a number of different destinations, seeing only key attractions and failing to get any 'real sense' (Ritzer, 1996: 76) of what a place has to offer. For Ritzer, such tourism is essentially superficial and instantiates the standardising calculability of package tours. It was not until his co-authored publication with Allan Liska in 1997 (also published as a separate essay in Ritzer's single authored text *Explorations and Extensions*) on the notions of

'McDisneyisation' and the 'post-tourist' that the rationalisation of tourism was treated to more concerted analysis.

In this later work, Ritzer examines tourism from the viewpoints of modernity and postmodernity, not as epochs indexing different social forms which can be 'read off' the historical development of society, but as *alternative perspectives* for examining social phenomena. He therefore takes 'modern' and 'postmodern' perspectives in turn to see what light they might shed on tourism and whether there is any incompatibility between them. With regard to the 'modern' perspective, Ritzer suggests that tourism has become increasingly subjected to the principles of McDonaldisation and that Disney theme parks are the industry's paradigm for this process. Following the works of Rojek (1993) and Bryman (1995), he uses the term McDisneyisation to allude to the difficulty of escaping the kinds of predictable and standardised holiday options of today. Of course, this view is very much contradictory to the idea that contemporary tourism is characterised by a diverse choice of holiday options.

Moving to the 'postmodern perspective', Ritzer sees compatibilities between the modern process of McDisneyisation and the postmodern notion of the 'post-tourist' (Feifer, 1985), that is the playful subject that accepts the commodification of tourism and is fully aware of the lack of possibility of an 'authentic' experience. Drawing upon Baudrillard, Ritzer argues that this post-tourist plays out her cynicism within an increasingly simulated tourist environment where her is nothing authentic left to be experienced – only copies of copies, and play with decontextualised signs. He uses this perspective to contradict MacCannell's idea that tourists look for authenticity when they travel. It seems to Ritzer that tourists, aware of the impossibility of the authentic, are in fact on the search for the inauthentic. Presumably the complementarity in these modern and postmodern perspectives on tourism lies in the fact that the rationalisation process is responsible for both views of the world.

Ritzer's complementary views of tourism as both rationalisation and simulacrum (copies of copies) speak of the irresistible encroachment of the *logic* of consumer capitalism into previously 'untouched' territories. This is a more expressly cultural, ideological account of tourism than the neo-Marxist one from earlier which takes relations of production as the sole driver of tourism. As the tourism industry commodifies yet more spaces into places, or perhaps more correctly, destinations for consumption through a process of rationalisation, local cultures and their 'authentic' ways of life have purportedly been transformed into spectacles for the tourist and in the process become 'eroded'.

According to this particular narrative of the impingement of capitalist exchange relations into 'traditional' ways of life, epistemic violence (that is, the infraction of indigenous knowledge systems) is being committed as the ideological and economic interests of modern societies threaten the local interests of the host society. A cursory look at the culture contacts literature suggests that such exchanges are socially, environmentally and culturally polluting for the Other, involving the distortion of the natural, indigenous, untainted values which purportedly underpin noncapitalist exchange relations in traditional societies with those of the nasty modern market economy. We return to this problematic division of modern and nonmodern exchanges later.

At the heart of both these corrosive grand narratives of modernity (the neo-Marxist and the neo-Weberian ones) is a temporarily framed tale of 'loss'. As Lyon and Colquhoun (1999: 192) argue, the use of the 'past as a leisure resource[,] ... a consciously therapeutic dimension to our lives' represents one response to the demands and challenges of our seemingly fast-moving times. This is an important point as it tells us something about the role and experience of time in tourism, and the kinds of trajectory along which tourist exchange, and the idea of the authentic Other, are frequently articulated. With regard to the authentic Other for instance, the role of time is clear. The Other is part of a premodern society, belonging to a nonindustrialised period where time is constructed and experienced differently from the time associated with modern society and its labour processes. Just think of modern advertisements telling us to relax and enjoy the slower pace of life, and the different rhythm and time of another culture. As Wang (2000: 97) so clearly points out: 'Modernity brings about not only a new temporal orders (schedules, routinization, pace of life and collective rhythm, etc.), but also a new consciousness of time'. Part of tourism's appeal, then, is that it sells a different kind of time – a slower pace of life – associated with nonmodern societies, which have not commodified their labour.

Lyon *et al*. identify a key contradiction of contemporary society as lying within the fact that many systems and structures limit our 'control' over time (e.g. the labour process), exacerbating our sense that it is always running out, while also promoting the illusion that we are masters of our own lives in this regard. They claim that we seek to maximise our psychological experience of time in terms of quantity *and* quality, to alleviate the fragmentation and uncertainty of contemporary experience in the West, by looking to the past (Brewis & Jack, 2005). They borrow especially from Toffler's discussion of alienation as a response to the 'future shock' induced by 'too much change in too short a time' (Toffler, cited in Lyon *et al.*, 2000: 19) in this regard. They do suggest, however, that most of us utilise the past only episodically and then as a coping mechanism (p. 20).

We could argue that this might be one of late Modernity's ambivalences, key to Wang's (2000) sociological thesis on tourism and modernity. But perhaps more than this is a wider concern that although contemporary life is perceived to be speeding up (we are constantly running out of time), it is doing so along the same capitalist economic and social trajectory that it always has been. In other words, speed does not mean social change; it just means a rhythmic intensification of the same. In this temporal sense, tourism does not in fact offer anything different – it simply carries us along the same trajectory whether by means of the alienating labour process, or the various structures and practices of rationalisation. Tourism as exchange might often be viewed as proceeding along these lines, as structure and action, either of alienation or rationalisation.

The Problem with Big Pictures

Such framings of the relationship between exchange and authenticity under Modernity are problematic from a number of perspectives. For one, they both restrict our view of tourism as exchange to the marketplace and more importantly to that of commodified relations of production and/or consumption. In this regard, exchange becomes part of a highly abstracted, decontextualised notion of a market disconnected from the specifics of social and cultural context. For exchange, read 'commodity exchange', achieved on the basis of money and a system where all objects are rendered equivalent by the process of commodification.

Second, both neo-Marxist and neo-Weberian framings of tourist exchange confine it to an act of consumption that involves the passive assimilation of dominant ideology. Whether part of the alienating conditions of labour, or the rationalising conditions for consumption, such grand narratives of history and progress homogenise individual actions, intentions and beliefs, and implicitly reduce them to a passive 'taking-up' of the interests of the owners of capital. Consumption is somewhat monolithic here, bereft of the colour and diversity of human actions and the possibility of resistance to such structures of power. Consumers are reduced to dupes of the inter-relation of the economic and cultural spheres. As Desmond (2003: 39) asks .'(...) how might (McDonaldization) explain the irruption of countercultural forces such as punk rock, rave culture and animal rights activism which sprang from the grey "McDonaldized loins of suburbia?'". As de Certeau (1984) reminds us, we need to consider consumption as a practice of everyday life, one where consumers' uses of objects form part of sometimes highly subversive relations to capitalist production.

Third, as largely dystopic tales of Modernity, these grand narratives presuppose and work in different ways with the theme of nostalgia, a harking back to the halcyon days of integrated and 'whole' social relations that have now been lost. According to Parker (2002), there is much that is elitist in such views. He argues that Ritzer's view of McDonalds is disdainful and patronising of those who consume fast and convenience foods. It is also presumptive of a 'better' past, 'an older, slower, quieter world' (Parker, 2002: 32), which may well never have existed, but is useful in the creation and legitimation of stories about the corrosive nature of the present. Themes of utopia and dystopia loom sometimes overly large in such stories.

There is of course a long tradition of elitism in tourism research, a tradition that mirrors the privilege of tourism itself and is fed by Romanticism (Boorstin, 1961; Buzard, 1993; Fussell, 1980; MacCannell, 1976). The neo-Marxian analyses that attempt to understand tourism as a set of alienated social relations may be understood as a reaction to this tradition and as an attempt to wrench tourism into other, less elitist, frameworks. What we end up with here, as with other macro and micro emphases in the tourism literature, is a situation in which there is a danger of creating binaries, out of such dialectical movements, between the grand and the intimate, the Romantic and the Realist, the material and the sentimental.

Fourth, and related to this latter point, in terms of thinking about the act of tourism itself, such narratives provide a limiting cartographical view of tourist movement. This is important if, as Meethan (2001) argues, the defining feature of tourism is the commodification of space. In this regard, Lury (1997: 75) perceptively suggests that:

Tourism has (...) been considered in terms of people travelling to places, or perhaps more specifically people travelling to places as cultures in mapped space. There is in this approach a presumption of not only a unity of place and culture, but also of an immobility of both in relation to a fixed, cartographically coordinated space, with the tourist as one of the wandering figures whose travels, para-doxically, fix places and cultures in this ordered space.

This urge to fix places and cultures serves not only to reify and abstract place but also to homogenise the cultural identities of those living within its boundaries. The tourist of Western industrialised society is supposedly part of a wider homogeneous group of universal workers alienated by the conditions of the labour process. They seek solace in the Other, where the Other is viewed as an equally homogeneous, but diametrically opposed set of people that are untainted by the violent orderings of Modernity. This is a key issue, for tourism studies relies on us and them; on the dominant flows between the West and the rest (Alneng, 2002). Attempting to break with this, and with other dominant ways of seeing the tourist world, is a key challenge for our work in this book. As such it has led us, for instance, to take flows along the west–west axis, tourism amongst industrialised, late capitalist, western subjects, as the focus for our study.

Of course, our choice to focus on the example of neo-Marxian and neo-Weberian grand narratives in this chapter is partial and hardly representative of all the nuanced ways in which tourism studies might approach the study of exchange. However, it does seem to us that these have been and continue to be central theoretical frameworks and working traditions in tourism studies. Add to these the other binarising and dialectical tendencies we have identified thus far and we begin to see how we may become bound by the very intellectual traditions, literatures, methodologies and disciplinary practices that sustain us and our field. The profound influence of French poststructuralism across the Arts, Humanities and Social Sciences is a case in point.

We are not advocating a break with these traditions, quite the opposite. But we do wish to argue both for an interdisciplinarity that, whilst respecting the integrity of tradition, takes the different disciplinary influences in tourism studies to their outer edges in an attempt to live with and through tourist phenomena. In other words, we are attempting to *begin* with phenomena, be they baggage, books, the excitement of certain encounters with other people, the fridge in a youth hostel, the memories awakened by being tourists and doing research. Here we take our cue from Merleau-Ponty, David Abram and Tim Ingold (Abram, 1997; Ingold, 2000; Merleau-Ponty, 2002) in attempts to understand exchange in tourism as a form of perceptual reciprocity, that engages the senses, the material world, that triggers the emotions and stimulates the intellect and that does so in ways that shift, share and change the balance and nature of power and of social and cultural practices.

If we look outside the dominant traditions in and of tourism to other disciplines, we see alternatives that help not only to sustain a critique of them, but also to provide some conceptual alternatives. In anthropology and material culture, for instance, the notion of exchange has been a central conceptual vehicle for the investigation of society and culture. It has much to offer in terms of a more nuanced understanding of exchange. And in recent times, cultural studies and cultural geography have witnessed a flurry of publications on the notion of cultural economy that explicitly addresses the relationship between the spheres of culture and those of the economy. It is to anthropology that we move next.

Anthropology and Exchange

In turning to anthropology, we try not only to extend a kind of 'internal' critique of the grand narratives presented earlier, but also to offer a different version of the nature and function of exchange. It is clear from the anthropological literature that the whole category of exchange, and its positioning of human relations and subjectivities, can be understood in radically different ways. And just as there are as many versions of anthropology as there are anthropologists, so too there are many versions of exchange.

For one, research in anthropology allows us to directly and critically address the manner in which much sociology has tended to elevate capitalist principles of exchange and their relations to a highly abstracted marketplace based on forms of equivalence. According to Eriksen (1995), one of the effects of such a privileged view of market exchange is that it tends to separate out the economy too much as an institution. There are a number of ramifications of such a 'separating out'. The first is that it either renders subservient or even invisible other (i.e. principally nonmarket, noncapitalist) principles and practices of exchange. In addressing why there exists in industrialised (and especially consumer) societies the myth that we exchange 'more' than in other kinds of society, John Davis (1992: 2) argues:

> One reason we think we exchange more than some other peoples do is that we overemphasize the part which commerce plays in our lives. But it is rather a crude stereotype of our economies to think that they are overwhelmingly dominated by commerce and the state. It is rather easy to ignore all our exchanges which we make with friends and relatives rather than in a market or shop: we define them as trivial, concerned with small quantities, below the threshold of perception, even though we know from our own experience that they are politically charged, have considerable symbolic importance,

and can have consequences for material as well as emotional, spiritual well-being.

As Davis makes clear, and as alluded to in the introduction to this chapter, there is a tendency, and perhaps not a surprising one in (post)modern Western industrial societies, to *overemphasise* commercial exchange as an economic form. It therefore would be instructive to consider noncapitalist or noncommercial forms of exchange, in so far as any 'rigid' distinction between commercial and noncommercial spaces can be maintained in a world of intensely global capital. Perhaps more specifically, we share the concerns of Dant (2000) who argues that the notion of exchange has been excessively focused around the cash nexus, a fact that has unduly circumscribed our understandings of material culture. Outwith this cash nexus, the work of Davis (1992) who outlines a multiplicity of different 'repertoires of exchange', from arbitrage and banking to charity and robbery, in the context of the UK (see p. 92 in particular) and the work of Gregson and Crewe (1997) on car boot sales, are excellent examples of studies addressing exchange systems outwith the abstracted capitalist market.

The tradition of the gift, and the notion of gift economies, has often been the focus for anthropologists (most notably Mauss, 1990) in considering alternative forms of exchange to those in 'the West'. Indeed, the contrastive differentiation of 'commodity exchange' and 'gift exchange' has provided a popular way of articulating the question of alternatives to the market. We do not intend to dwell at length on the debates surrounding the meaning, functions and empirical accuracy of studies of the gift (this has been done most comprehensively elsewhere). Rather, what is of key interest to us is that gift exchange is supposed to based on a completely different rationale to that of commodity exchange, and that historically, gift exchange has been notionally confined to nonmodern societies such as the Trobriand Islands, the site of Malinowski's study of the Massim peoples (Malinowski, 1922). Desmond (2003) neatly summarises some of the key differences between gift and commodity exchange. His work is presented in Box 2.

As Desmond's useful comparative work demonstrates, the practice of the gift exchange is one that enacts different values and takes different forms to that of the commodity exchange. It is more about reciprocity and the creation and reproduction of social bonds and community through embodied social relations, than it is about individual gain and the use of an abstracted market for social relations. They are characterised by different kinds of rationality and they reproduce different ideas of society.

Box 2 Summary of principal differences between gift exchange and commodity exchange

---

Gift exchange
Status increases as one gives things away
Objects are inalienable
Establishes a qualitative relation between parties, builds community
The exchange is backed by the need to reciprocate
The exchange creates dependence
The exchange creates worth
The decision process is based on the demands of community and reciprocity

Commodity exchange
Status increases as one accumulates things
Objects are alienable
Establishes quantitative relations between strangers
The exchange is backed by law
The exchange maintains independence
The exchange creates values
The process is rational, calculative and linear

Source: Desmond, 2003: 147

However, Davis (1992) and Eriksen (1995) strongly argue against the tendency to place forms of exchange into either West (with commodity exchange) or the rest (with gift exchange). As we pointed out earlier, there has been a tendency to reduce our understandings of forms of exchange and the economy through the use of binary oppositions. In anthropology notably, one of the most enduring of these has been that the West exhibits capitalist exchange relations and the rest have gift economies, or at least those based on principles of reciprocity.

In fact this distinction is too clear-cut and it is increasingly unusual to find 'pure' examples of either type of exchange relation. Indeed much anthropological study has looked at the way in which the cultural logics of so-called traditional societies have been affected by the introduction of money-based exchange. A famous example of a study documenting the impact of the introduction of Western money into a previously nonmonetary economy is Bohannan's (1955, 1959) work on the Tiv of northern Nigeria.

The second ramification of the separating out of the economy as an institution, of which Eriksen (1995) talks, is that it often results in the notion and practice of exchange being stripped of the context in which it takes place. We would argue that one cannot divorce the study of exchange, whether it be gift or commodity, exchange, from its social and cultural context. The difficulty here seems to be that money-based exchange, and the abstracted market, is seen to be a naturally occurring phenomenon that somehow exists outside of social and cultural systems. The use of money is taken as something that 'just happens' and 'just is'. As the edited anthropological work of Parry and Bloch (1989) more than adequately demonstrates, however, money is not an a-social or a-cultural matter. Money is a profoundly symbolic phenomenon that shapes and is shaped by its social and cultural environment, and which as such has important moral qualities to it.

Expanding on this theme of the inseparability of the economic and cultural spheres, recent times have witnessed a mushrooming of publications addressing their interface. In their recent book on cultural economy, du Gay and Pryke (2002) for example suggest that the economic and cultural spheres, whilst not reducible to each other, are mutually constitutive of each other. Peter Jackson's recent co-edited text on *Commercial Cultures* specifically sets out to 'transcend the dualism' between common-sense understandings of commerce and of culture. To quote them:

> (...) the commitment made by this book .... is to find a way to transcend that dualism and to explore the way that various aspects of cultural production – in fashion, publishing and retailing among other sectors – are inherently concerned with the commodification of various kinds of cultural difference (...). Conversely, we aim to show how the apparently rational calculus of the market is inescapably embedded in a range of cultural process. We are not seeking to 'reduce' the cultural to the economic (or vice versa), or to show that either side of the equation is more significant than the other. Rather, we assume that the world is full of what might be called the authentic hybrid that is commercial cultures. (Jackson *et al.*, 2000: 1)

In the fourth and final subsection of this chapter, we turn to work that helps inform and create a framework that develops the points mentioned above. It combines a framework for exchange based on Daniel Miller's (2000) 'birth of values project' with recent critical work on material culture, most notably by Celia Lury (1996, 1997).

A Material–Cultural Approach to Exchange

In common with the criticisms outlined earlier, Daniel Miller (2000) is of the view that the analysis of exchange in anthropology, as well as other disciplines, has tended to be based·on rather simple dualisms. He makes particular note of '[the] rather simplistic contrast between two idealised extremes – the "gift" and "commodities" (or sometimes the "market"), and the two distinct ways these are thought to constitute value' (Miller, 2000: 77). Miller argues that the long-standing debate between these two forms of exchange might well have run its course because the opposition can be so easily turned on its head. In its stead, he suggests an alternative trajectory for studying exchange, which might have the capacity to transcend such a facile and easily disrupted dualism. His idea is that researchers should focus their attention at the level of cultural values and trace their changing forms and effects through different and sometimes transformative contexts. This could, for instance, involve tracing the values articulated around particular material objects in commodity systems of exchange and how these values become transformed as the objects move into alternative repertoires of exchange. Perhaps then we might view tourism as an 'exchange of values', looking at how, where and when different kinds of value are exchanged and what their consequences are.

As alluded to above, the negotiation and transformation of values is enacted as part of a wider set of object-relations, and it is these object-relations that have formed the basis of the now well developed literature on material culture. In her text on *Consumer Culture*, Celia Lury (1996) neatly summarises two interconnected approaches to the study of material culture as part of her wider thesis on the development of consumer cultures in Euro-American societies. On the one hand, it is pertinent to explore the manner in which social lives have things, and the role of objects in the construction of social identities (as part of which we might include the tourist identity). On the other, we might also consider that an object itself has a social life and therefore a biography of uses and meanings. This latter sense of object-relations has stimulated much of the recent debate in this area, especially around the question of the relation between the material and the cultural, and whether or not the concept of culture constitutes a 'useful' vehicle for social analysis. These debates enable us to develop Miller's call for an emphasis on the transformative nature of values in consonance with recent debates on material culture.

To begin with, we are influenced by the work of Appadurai (1986), who succeeded in shifting the focus in anthropology from the forms and functions of exchange to the things that are exchanged, thus '[making] it possible to argue that what creates the link between exchange and value is *politics*, construed broadly' (Appadurai, 1986: 3). Exchange therefore becomes a political activity for our purposes, intimately associated with the past and the future and a range of values, bedded in social systems, complex and diffuse and made manifest in things. This aspect of the object allows us to consider what Kopytoff (1986: 66–67) terms the 'cultural biography of the object':

> In doing the biography of a thing, one would ask questions similar to those one asks about people: What, sociologically, are the biographical possibilities inherent in its 'status' and in the period and culture, and how are these possibilities realized? Where does the thing come from and who made it? What has been its career so far, and what do people consider to be an ideal career for such things? What are the recognized 'ages' or periods in the thing's life, and what are the cultural markers for them: How does the thing's use change with its age, and what happens to it when it reaches the end of its usefulness? [...] Biographies of things can make salient what might otherwise remain obscure.

In writing selected biographies of the material objects in our travel bags, those that make a particular claim upon us, we hope to show something of the politics and values that find their way into equally political and value laden bags. Asking such questions of culture can take us into a fascinating realm of cultural history, historicity, propaganda, nation building and deconstruction. But equally it takes us into relationships, sentiment, emotional fulfilment, the satisfaction of knowing things to be in their right place. Into the things we live for as well as the things we live by (Eagleton, 2000). It allows us to observe something of the role played by selected cultural artefacts in the politics of culture, and it enables us to begin to add something distinctive to the dynamics of alterity.

Material culture has long been a focus for anthropologists. Appadurai's work is, after all, already dated, despite its now classic textbook status. One of the prime ways of focusing on material culture was, until recently, to concentrate on the action of exchange rather than on the objects themselves. The shift to the focus on the object is both physical and linguistic/metaphorical. Objects wrapped in actions become symbolic, as do actions wrapped in objects. The more recent attention to material culture and to the coinages 'materiality' or 'material life' is worth comment, however.

The issue is not merely that focus on material culture enables the showing or the physical manifestation of politics and values, but that the very shifting focus in social science to the material world is part of a potentially more significant theoretical move. The focus on language and culture during the last three decades, since the [in]famous linguistic and cultural turns and the forays into poststructuralism, has allowed attention to be paid to the processes of cultural construction, the role of language and text in creating culture and to the slippery nature of representation. But all the time, material life continued in almost unnoticed, intimate ways, piling, accumulating, burning, breaking, moulding and shaping, giving and receiving. It is rather like digging away in Foucault's earth to find, first and foremost, an artefact, as opposed to power, discourse, process or punishment and, as this long forgotten thing is carefully dusted down, restored and polished, it begins to show its usefulness again. Returning to material culture at this juncture is thus like a trip to the attic, bringing down dusty boxes and being excited by their long lost potential or by their sheer curiosity value. In this context then, 'materiality' and 'material life' are perhaps more interesting terms for us than 'material culture'. After all, culture is now such a written, unwritten, rewritten term that it has been declared both impractical and even moribund, requiring at the very least a stint in the attic, according to several leading cultural theorists, including Terry Eagleton and Tim Ingold. Ingold (1993: 230) states that:

> It could be said, I suppose, that through the deployment of the concept of culture anthropology has created the problem of translation rather than solved it. Having divided the world, through an operation of inversion, we are now left with the pieces that have to be connected together again through translation. Would it not be preferable to move in the opposite direction, to recover the foundational continuity, and from that basis to challenge the hegemony of an alienating discourse? If so, then the concept of culture, as a key term of that discourse, will have to go.

Ingold's view here is located within the debates around the importance and uniqueness of cultures and of identity construction in anthropology. His attempts to flatten the sides of the cultural box have a specific resonance within anthropology; his understanding of translation is metaphorical, not linguistic, and his 'continuous world' is one where Babel has no place. Nonetheless, his view of the problems caused by parcelling the world up into neat cultural packages is helpful in at least breaking something of the hegemony of the discourse on culture.

Eagleton (2000) also has problems with the idea of culture, but his problems are not so much concerned with the way the culture concept divides up the world, but with the political and material aspects of life that a dominant focus on culture obscures. In this respect he shows himself to be continuing much of the work of Raymond Williams:

Culture is not only what we live by. It is also, in great measure, what we live for. [...] We have seen how culture has assumed a new political importance. But it has grown at the same time immodest and overweening. It is time, while acknowledging its significance, to put it back in its place. (Eagleton, 2000: 131)

For Eagleton, then, our focus should shift from the preoccupation with culture that he sees as characterising the modern age, to the material issues that characterise the politics of culture, rather than cultural politics.

References to material life, materiality and to links or relationships between the action of exchange, things and values point, however, to more than just a 'material turn' in social theory. Schiffer and Miller (1999), as well as carefully documenting the marginal status of material-culture studies in psychology, sociology and anthropology, point to the way that understanding the relationship between people and artefact as secondary to the construction processes of culture merely repeats conventional ontology in an empirical domain. The key studies of Appadurai (1986), Miller (1999) and Douglas and Isherwood (1978) are all singled out for critique in this way and their use of the term material culture is seen to be problematic:

Even the term material culture subordinates artefacts to a cultural frame of reference, acknowledging objects but denying the materiality of human life. Apparently, material-culture studies have been domesticated and pose little threat to theoretical hegemonies, much less traditional ontology. (Schiffer & Miller, 1999: 6)

The 'material turn' and the return to questions of ontology – that is to say how we understand our being in the world, rather than how we construct our knowledge of the world – may indeed be a reaction to the focus on construction, language and culture. This is not, however, in any way to say that the work undertaken to understand human worlds as culturally constructed and to understand the primary, often sensuous role, played by language in the processes of representation is invalid and rendered useless by a change in focus or direction. Indeed precisely the turn towards materiality can imbue the debates regarding cultural construction and the linguistic process with new dimensions, for they enable us to focus anew on the physical dimensions of language and construction.

Lury's (1997) recent work on travelling cultures would seem to echo some of these suspicions about the 'cultural' in material culture, or perhaps more specifically the privileged focus on the human rather than the object of human-object relations. As highlighted earlier, Lury is critical of the fixed cartographies often associated with tourism, i.e. the notion that tourists move from one clearly mapped and fixed place through travel to another clearly mapped and fixed place, from which they may return with some kind of object for memory-formation. Drawing upon the work of anthropologist James Clifford (1992), Lury critiques the fixity of opposition between travelling and dwelling which such cartographical organisation presupposes – that the tourist travels from one discrete place called home and then dwells in another discrete place called abroad. Her interest lies in the notion of 'travelling cultures', detached from particular places, which place a focus on the dynamics of dwelling/travelling. In terms of our current discussion on material culture, Lury importantly considers these travelling cultures not only in terms of the people that move in and with them, but in relation to the objects that also dwell and travel.

Lury argues that tourism research has failed to draw upon the kinds of academic interest in objects, especially in the discipline of anthropology, outlined earlier in this section. This she sees as problematic as it condemns objects to be seen as merely the 'extended baggage' (Lury, 1997: 76) of the traveller. Using Clifford's dwelling-travelling dynamic, she believes that objects comprise tourism not just because tourism involves objects-in-motion, but also because it involves objects that stay still. This dynamic of motion and rest is very much reflective of Clifford's description of travel. She does this to address what she calls the 'double omission' of tourism studies (p. 76), that is its inability to 'identify the significance of the travelling/dwelling relations of people and objects for the practices of tourism' (p. 76).

Our construction of a broader understanding of exchange draws centrally upon this work by Lury. Sections two and three set out to instantiate and concretise this interdependence of the travelling and dwelling of both people and objects as it was encountered and experienced in our study. What Lury's theoretical work offers us is a way of taking the role of objects within tourism seriously by demonstrating not only that they are centrally constitutive of its travelling cultures, but also that it can coexist happily with the emphasis on the negotiation of values outlined earlier in the work of Daniel Miller.

Whilst we are reluctant to condemn the culture concept too quickly to the attic (as Eagleton suggests we do), we are keen to offer a broadened conceptual frame for the study of tourist exchange within which material objects play an important role. We also see the phenomena of touristic material life, not as simple material objects, but as bound into sensory, participatory relations of exchange, triggering a range of thoughts, feelings, hues and, shades of emotions and memories in tourists' everyday lives.

In sum, this chapter has set out to consider how we might understand the phenomena of exchange as both actual tourist practice, and also as a metaphor for tourism itself. We began by examining the manner in which certain sociological grand narratives of tourism circumscribed the manner in which we might come to understand tourism as exchange. To reiterate, they tended to confine exchange to the principles of the decontextualised market and of capitalist relations of production. Tourism was either an outcome of the alienating or the rationalising impulses of modern society, or a recompense to them.

In seeking to broaden an approach to exchange, we engaged with some anthropological literature which demonstrated not only the highly embedded nature of tourist exchange, but also the possibilities of alternative structures and practices of exchange both within and outwith so-called 'modern' societies. Such insights impel us to investigate tourist exchange and tourism as exchange from the local viewpoint of tourism as cultural practice, rather than from the purview of the historicising grand narrative. From this perspective, it becomes germane to ask what other forms of exchange, apart from those of the commodified market, might exist within the tourist arena? We ask how non-money-based forms of exchange come to pass. How do they happen? What practices are involved and what kinds of objects are exchanged? And what are the wider ramifications of these intercultural relations?

The final part of this chapter investigated how contemporary debates on material culture might also contribute to our conceptualisation of exchange. In this respect, we brought together Daniel Miller's insistence on an investigation of changing values in relation to objects with Celia Lury's concern for the social life of things. Here we are courted to explore the exchange practices of tourism from within the travelling–dwelling relations of people and objects. Using both Miller and Lury, we are encouraged to attend not only to the cultural values attached to the exchange of tourist objects, but also to the materiality of the objects themselves, and their role in travelling cultures. In the next chapter we look at reflexivity as a key exchange practice in the travelling culture of our research relationship and explore it within a wider account of our research methods.

Chapter 3
Doing Being Tourists

The focus in our previous chapter was on developing a conceptual framework for the study of exchange in tourism contexts. In this chapter we turn to questions of method and methodology, and discuss not only the kinds of materials that have shaped our narrative of tourism in this book, but also the nature and form of the research process through which these materials were garnered. In other words, we focus on the 'what' and 'how' questions of knowledge construction, from which we tell our stories of the intercultural life of exchange.

When packing for the field we both packed bags for our data. We took boxes with us that contained materials already drawn together in the preparation process, boxes of books by Mary Douglas, James Clifford, John Urry and Walter Benjamin. We had brochures and guidebooks. We had tape-recorders, laptops, cameras and notebooks ready for inscription. We also had separate folders and bags for materials that we fully intended to collect on the way. To return home empty-handed would be to have failed.

In discussing how and with what we came to fill these empty bags, we pay particular attention to aspects of the dialogic process of collaboration that formed the basis of our ethnographic method. Here we comment on the importance of 'thin spaces' ('off stage' places of rest, writing and reflection) for the production of 'thick descriptions' and the patterning of field relations. Making visible these thin spaces at the margins of our ethnographic enterprise, provides a vehicle for clarifying the nature of our authorship of this piece, and for illuminating the material conditions of production for this work 'beyond' the projected field. In turn, we use this to offer sustained comment on the concept of reflexivity and its place in our methodology. Reflexivity we address not only as a response to its centrality in contemporary texts on methodology, but also, and perhaps more importantly, because it forms an embedded part of the everyday processes of social action.

Bags of Data: From Thin Spaces to Thick Description and Back

We could, of course, not have had any bags for data. We could have engaged with questions of tourism in Scotland without a projected 'field'.

35

We did not need to do participant observation, in tourist time, with other tourists. We did not need to listen to tourists tell tales, eat ice creams, drive Ford Focus hire cars, take pictures and embed their lives for a moment in time in a social and material space.

Historical research suggests itself. The holdings in the university and national libraries of Scotland provide plenty of materials and archives for tracing tourism through history. We could have done surveys, used questionnaires, recorded discourse, analysed representations in fiction, film and music. Where others have pursued their research questions in these forms we had work to rely on (Buzard, 1993; Macdonald, 1997). The questions we were asking, however, required a present time engagement with tourism as it happened. If, as Augé (1995: 18) suggests, 'anthropological research deals in the present with the question of the other', then it is closely allied in form, and in the premise of its object, to tourism itself – which also, we argue, deals in the present with the other. Ethnography, as anthropology's defining research method, therefore suggested itself.

An unusual, though by no means unique, aspect to this research, was its joint nature. The fieldwork was undertaken by two people, friends and colleagues. Doing the ethnography together resembled going on holiday together. Holidays are largely shared social experiences, whilst ethnography is largely a solitary pursuit. As a consequence, the two-way relation of ethnographer to field, became a three-way relation of ethnographers to fields, and to each other. The material presence of another human being engaged in the same action meant that narratives of the field occurred and were shared constantly in this milieu. There was not one, but two fields of engagement: the field of participant observation where we became 'tourist-ethnographers' and the equally present field of academic research.

With respect to the former, the *habitus*, what Bourdieu (2000) terms the *learned disposition for action*, of tourist-ethnographer was one that emerged through the practice of the tourist field. An interesting aspect of this was the manner in which we 'practised' tourist-ethnography through the projection, enactment and inhabitance of different kinds of research space. In addition to the construction of 'stages' of tourist action for description and interpretation, what we might consider the 'meat' of our body of work, we also dwelled in what might be termed 'thin spaces'. These were spaces which we sought out 'away from the crowds', in the non-places of anonymous cafés (Augé, 1995) or the non-place of empty nature where there was little danger of meeting tourist 'data'. These might be considered skeletal locations; fundamental to the support of our body of work, but largely invisible and rarely commented upon in texts.

These spaces, it should be noted, had three particular qualities. They were nourishing of the body, they were quiet and they were spacious enough to be free from distraction. This searching out of thin spaces – places where we rested rather than worked – this retreat 'from the field in the field', is akin to the return to the ethnographer's 'tent'. These private spaces are usually made invisible in ethnographic reports of the field, often deemed 'irrelevant' to the recording of empirical evidence, despite being the very spaces in which the actual recording occurs.

Our study sets out to bring some kind of visibility to these thin spaces of rest and recording, where we, notionally at least, switched off our 'researchers' on-switch. It is from the tent, also, that the life that is not recorded in the reflexive field notebook or ethnographer's journal, takes place: the writing home; the calls to loved-ones; the reassurance that 'work is progressing well and the trip has been worthwhile'; the sorting out at a distance of the practicalities of physical absence from home or work; the call to the bank; the conversations about the weather; the writing of postcards that usually do not figure; the sharing of the rough and ready ideas at a distance that enables those ideas to form or to fall. Walter Benjamin gives us some clues as to the importance of such thin spaces for the ethnographic enterprise.

Benjamin, in his writing on the import of Brechtian theatre and *Verfremdung* (distantiation), signals the need for a *Halt* – a pause in the flow and action of everyday life, or in the narrative flow of epic theatre. *Verfremdung* is allied to reflexivity. The rupture of action by, in our case, the making present of what is otherwise absent – home – enables the generation of stories, tales from the road, reflections on the quality of experiences. There is little material difference between calls home from the field and calls home from holiday. We were, in our public phone boxes, privy to snatches of many other such conversations. Without the pause and the rupture, and the attending to other concerns, there can be no reflexivity. In the youth hostels at night we, along with our fellow tourists, were busy at times, attending to home, writing postcards and letters, interrupting the experience of the present with a practical attention to past and future. We shall return to these aspects of tourism in Part 2. Addressing these thin spaces, and their importance in terms of rupturing and bringing awareness to our routines, underlines the need to account for the nature and material forms of reflexivity in our research.

Our contention and consequent focus in this chapter, then, concerns the centrality of the material life of reflexivity to the work of ethnography and the work of tourism. We focus on the felt-experiences, the sensuous and the emotional work that attends reflexivity in research contexts. Just as the material relations that enable tourism to occur matter, so too do those that enabled our work of reflexivity. In *Pascalian Meditations* (2000), Bourdieu makes it clear that the freedom from material necessity is a precondition of the scholastic disposition. Reflexivity is important, in Bourdieusian terms, precisely because it refuses to take for granted the same set of social and economic conditions over the sustained period of research, that enable the temporary 'leisure' of tourism.

A Material Life of Reflexivity

What our work of exchange basically entailed was joint travel, joint research and the creation of joint (in addition to individual) texts. What we outline therefore in this chapter is an emergent methodological form, which is constructed and experienced through the multiple, reflexive exchanges between researchers, participants and the symbolic and material spaces that they come to enact. Our emphasis therefore is on methodology as a *process*, both personal and social, which comes to be written (Clifford & Marcus, 1986) but also felt reflexively. Exchange and dialogue lay at the heart of our methodological practice not just in an attempt to mirror the substantive concern of our research, but also to act as a vehicle for the recording and subsequent scrutiny of our reflexive practices. Acts of exchange are embodied social and reflexive action. They have emotional resonance. To discuss exchange and to discuss material life is to attend to the conditions in which knowledge is constructed as a casuistic necessity in the process of translating critical, cultural theories of tourism study into research *praxis*.

These conditions may appear to concern epistemological questions. However, at the end of the day, when the sun is setting over the bay, when the notebook finally closes, and conversation turns again to its richness, it is in thin places of the day's experience that we begin to be more friends than colleagues, more companions than tourists meeting in the same place for a day. We insert a pause not for sleep, but for the conversations we do not record. As questions about the construction of our knowledge fade and our concerns turn to the conditions of our existence, we experience our research differently.

We are given an opportunity for a rupture, a pause, and for being present and for *feeling* the day that has passed, the lives that we live and for imagining the work to come. There is a paradox at work in these thin spaces, the spaces we do not normally record, for they are thick with potential. These are the times when we can give ourselves over to a more dream-like sensing of experience, allowing what Brueggemann terms 'the kind of at-homeness that precludes hostility, competition, avarice and insecurity' (Brueggemann, 1999: 50). Reflexivity here is quite the opposite of a pause for thought. It is indeed a pause, but one which sees us resisting the abundant claims of the tourist phenomena of the day and attending to other claims. Far from being the times in the day when we stop being reflexive, these are the times when we are abundantly claimed by reflexivity. Let us now clarify the sense in which we have come to understand the term *reflexivity*.

Reflexivity

The concept of reflexivity indexes social scientific concerns for the sociocultural and political conditions in which all forms of knowledge (academic and lay) are constructed. Specifically it represents the belief that the conditions in which claims to knowledge are made are centrally constitutive of these claims. In his notion of discourse, for example, Foucault (1978) argues that the concepts, language and social practices which are brought to bear in constructing knowledge (that is discursive practices) serve to define, delimit and thus mask out alternative narratives of truth and knowledge. As such, these conceptual and linguistic practices are the conditions that serve to construct claims to truth. Cognisance of such acts of truth-building is the stuff of reflexivity. But what exactly does reflexivity look like? What form or forms does it take? Is reflexivity a stream of consciousness through which transcendence to some kind of 'pure', 'conditionless' knowledge might be achieved? What might be its active material base, in, for instance, the specific context of our fieldwork?

Conventionally reflexivity seems to be taken as an act or activity that is performed on data, *after* it has been constructed through various theoretical narratives. Kind of like having a one-minute silence to reflect, nae mourn the docile bodies on which social scientific texts are written. Trying to be close to the corpses of academic butchery once we have fled the scene of the crime. Some kind of postcoital epistemological cigarette? Concerns might be raised about the temporal veracity of such reflexive claims. As both Krippendorff (1994) and, in particular, May (1998) are at great pains to point out, reflexivity is *not* an act performed on data in an attempt to capture and 'delete' its political contingencies. Reflexivity is an integral part of the practices in which we engage during social action. That is to say, reflexivity is already part of the social lives in which we engage. Based on this assertion, May outlines two distinct ways of understanding reflexivity. Given the importance of the concept of reflexivity to this work, it is worth quoting May (1998: 157–158) at length:

> Endogenous reflexivity refers to the ways in which the actions of members of a given community contribute to social reality itself. This encompasses not only an understanding of how interpretations and actions within the lifeworld contribute to the constitution of social reality, but also those same aspects within the social scientific community itself. Endogenous reflexivity thus relates to the knowledge we have of our immediate social and cultural milieux.[...] To this we have to add the dimension of referential reflexivity. This refers to the consequences which arise from a meeting between the reflexivity exhibited by actors within the social world and that exhibited by the researcher as part of the social scientific community. Referential reflexivity thus refers to the knowledge which is gained via an encounter with ways of life and ways of viewing the social world that are different from our own. The constitution of this discursive knowledge and its implication for social practice, enables an understanding of the conditions under which action takes place.

In other words, one form of reflexivity is found in the things people do without thinking, habitually. This is the kind of action Lévi-Strauss famously details in *La pensée sauvage* (Lévi-Strauss, 1962). Another, more rarefied, scholastic form of reflexivity is found in the things academics do to their objects. Both forms, in our view, may occur simultaneously. They are not neatly separated in time and space. This distinction is not dissimilar to Barthes' (1970, 1975) positing of the readerly and the writerly as differentiated modes of engagement with texts and with the act of reading, as we shall see later (Part 2).

To give an example. Tourists and ethnographers of tourism need public toilets. Once inside said public convenience, however, the ethnographer finds a whole world of potential for reflexivity. In these *thick* touristic spaces – spaces which claim us because they are rich in exchange, and in intercultural life – we come into contact with a clamour of texts, bodies and material objects that trigger feelings, reactions and exchange. There are notices in many languages concerning the disposal of waste and sanitary products, and we are up close with others, negotiating tight spaces, corners, sharing taps and dryers. For women at least, conversation often ensues. Answering a call of nature may be an action that is undertaken, in response to knowledge we have about the immediate social and cultural milieux'(there is a WC over there, so I can go and use it).

In practice, to repeat, these two senses of reflexivity are difficult to separate, but neglect of either is to mask further the discursive practices that serve to construct our knowledge. Reflexivity is therefore a concept which, it can be argued, sensitises us to the methodological challenges of translating the ontological and epistemological into research praxis. For us, the central challenge for the construction of a 'reflexive methodology' is being able to capture and subsequently render visible the primarily reflexive practices which serve to construct our knowledge of tourism. As such, the questions that concern us are *how* and *under what conditions* was our methodology constructed and how did our reflexive practices *take form*? We attempted to render visible, at least to some extent, our reflexive practices, the joint travel, by taping our conversation and exchanges. What follows is a discussion of these questions drawing on the transcriptions of our talk.

Reflexivity as Relational Form

Anthropological convention suggests that the ethnographer has been the central human (research) instrument, the key 'receptacle' for the collection of data during the process of fieldwork. In accordance with this convention, methodology can be regarded as an 'activity' which is experienced, enacted and embodied in a sovereign ethnographer who is dependent upon the field, upon the 'Other' for the provision of research material, whilst wholly dependent upon the privileged 'Self' for the subsequent construction of knowledge. Participant observation has traditionally provided the central method for the 'rooting' of such ethnographic methodology in the sovereign body of the ethnographer (Agar, 2000; Conquergood, 1991; Hammersley & Atkinson, 1983; Marcus, 1998).

In recent decades, the deconstruction of anthropology and the anthropological voice (Clifford, 1988; Clifford & Marcus, 1986) has led to differentiated ethnographic voices, and different modes of representing the evidence of the field (Clifford, 1988). Far from unproblematically assuming the stance of all-seeing, all-knowing observer-God, in control of meaning, ethnographers now carefully qualify their observations through dialogue, negotiation of meanings, insertion of their own positions and biases into their narratives and through different modes of representation. These draw not just on the classical ethnographic field report genre, but on museum pieces, poetry, photographs and multiple media that enable a differentiated engagement and a decentring of the ethnographer's position – at least in theory.

In practice, the ability to mobilise so many different modes and moods in writing further serves to establish the authority of the ethnographic voice, in an age which claims to privilege multivocality, and which gives cultural capital to those capable of mobilising evidence of their creative and reflexive practice. Distinction comes to those who make visible their reflexive practice, if not its materiality. Self-reflexivity is now credo for any ethnographer worth her salt. Thick description of the field is now thick and colourful, full of tone and hue and embellishment, all of which continue, through the conventions of problematised reflexivity, to add rhetorical force to ethnographic voices.

Our own ethnography sits comfortably, we think, with the now not quite so new ethnography. We too have pictures and textualising devices to hand to which we will have recourse throughout. Our intent in so doing is to bring the invisible practices of ethnography to the fore. Through our focus on reflexivity we are working, at least in this chapter, with thin descriptions, with grey areas of what is not part and parcel of the ethnographic life. That is to say, with the material conditions of reflexivity and relations beyond the field but which impact, nonetheless, on the field.

Having left the outskirts of Glasgow, we began our research in earnest in the car heading north to Fort William. Our transcripts document our concerns with the need to be thorough, to see spontaneous chatting and brief exchanges with tourists as 'interview' forms, to act like tourists and engage in tourist activities using tourist time, to reflect consistently on what we were doing and the ideas we were having. We wanted to let the structure of our work emerge as we got a 'rhythm' together, to use the metaphor of exchange to guide our methodology and importantly to be led and dictated in our activities by the tourists whom we encountered. The first example of us being led occurred on a Sunday, the day after we arrived on Skye, when Alison's parents, who were themselves fortuitously on the island on holiday gave us an insight into the tourist honeypots on the island, the different nationalities they had encountered and the different languages they had heard. They gave their reflections on the form and content of their intercultural exchanges, as reflexive tourists, with reflexive others:

Alison: So there weren't many foreigners there?
Mum: No, there weren't, but maybe they call there in the daytime.

Dad: Yeah you should go for coffee maybe, and think they'll call on the way in, because I think you can only get information there, about where to stay, but it's not. . .

Alison: But it's not an information centre.

Dad: It's an exhibition about Skye.

Mum: But as I said we've had three nice meals at the place in the main square. It has a very cosmopolitan clientele. And you go inside.

Dad: You get a nice fresh Skye salmon salad.

Mum: They close, last orders at seven, and their seven o'clock is getting on for ten to seven.

Dad: I would think, when the Germans have their coffee and *Kuchen* in the middle of the afternoon, that's when you want to go.

Mum: Yeah, I would think that would possibly be the time for that.

Alison: That's interesting.

Mum: We had a very good meal there.

Alison: We could go there when we're staying at either of the places up there.

Mum: But don't be put off by the menu card, they have a little thing that says today's specials and if there's nothing written on that outside, they've got it inside, but we've had two wonderful salmon salads there, excellent, but certainly there were a lot of continentals in there.

Even in this initial set of exchanges with Alison's parents, there is a vital point to be made about the 'location' of methodology. In short, it was not based on our sovereignty as individual researchers but in our relationality and mediated exchanges. It is about being the children of parents, about the way some of our choices bring us back to beginnings again, about the intense embarrassment that any child feels when her parents pop up in professional life. Whilst we were engaged in hard intellectual work, mum and dad had been on a boat trip, and had been 'tourist-watching' themselves, because of our somewhat 'jammy', in their view, research project. And this encounter lent a particular dimension to our future exchanges as researchers. Gavin now knew Alison's parents as people, capable of sitting out on terraces, drinking tea and engaging, respectfully, in conversation. When stories of Alison's holidays as a child filled their conversation later in their work together, the idea of Alison's parents needed no further fleshing out. And salmon salads were consumed with memories of these conversations.

It is nonetheless uncomfortable, and feels transgressive to bring this element of the research into view. It would be easier and equally possible to write a research report that did not mention meeting up with Alison's parents, that did not acknowledge any role they may have played in the development of ideas, the construction of the project, the emergence of itineraries or the positing of theses. Parents do not belong in ethnographic narratives, unless as the kin of those who are the object of study, but certainly not as parents of the ethnographer. That private lives interweave socially with our fields of research is rarely the focus of reflection on how data is gathered. Yet the enormous emotional investment in domestic relations has a significant bearing on the psychodynamics of our data and its orderings; relations we observe in others as ethnographers but excise so often in ourselves in order to retain a 'pure' ethnographic voice.

So, we were being led. As such, even at this early stage of the research, it can be suggested that the methodological imperative of engaging in 'participant observation' does not adequately encapsulate what we were actually doing when we lived the 'ethnographic life' (Rose, 1990). Rose believes that a strict division between the life of the field and the life of the researcher and a tight schedule and planning of ethnographic research cannot reach the heart of the researched culture. In this regard, our reflexive practices, the sense which we are making of Alison's parents, straddled more traditional delineations of methodological theory and practice as each became submerged into the other through engagement in field relations. We listened together to each other's stories and then, over time, allowed the subsequent settling of these experiences and memories into the different and separate spaces of our lives, into different yet similar stories, and shared memories. Our work of exchange as *praxis* is thus grounded in the relativities, not the sovereign *cogito* of the researcher.

Such relativities cut both ways, however. As part of this, we needed to be recorded by other people, to be acknowledged by the field (Augé, 1995), as much as we needed to acknowledge them. Unless this happened, our ethnography seemed fruitless. As Rose explains in his calls for a 'radical ethnography' (1990: 45): 'It is not in criticism, but in relationships that the promise of the field resides [...] relationships across boundaries are more important than methodology per se.' The point that Rose makes is vital as, in many ways, it takes the methodological space of the ethnographer away from notions of his or her integrated body as the ethnographic human instrument to that of the embodied interspace, to what Buber terms the 'dialogical' (Buber, 1954) and to Ingold's notions of an interagentivity (Ingold, 2000). A particular shade of this relational form was narrative.

Reflexivity as Narrative Form

On the Sunday afternoon, just one day after arriving on Skye, we were buzzing, excited by the tourist tales narrated to us by Alison's parents. We were already 'making our world go round by chasing our tales' (Reid, 1992), unexpectedly and seemingly prematurely overwhelmed by Skye, the claims the work was making on us and the sheer excitement of it all. We found our talk and our feelings to be tormented by the *need for form*, the need to bring structure to the seeming uncontrollability of the exchanges that we had been experiencing.

Form may not be familiar or tidy. It may be frustrating, disordered, confused, ambiguous, but have shape nonetheless. The form of this book seeks to reflect something of the form of our ethnography. 'Every element of form has an active material base' says Raymond Williams (1977). What comes out of the material and intellectual travel bag and is used depends on the unexpected and unknowable nature of ethnographic encounter, of exchange and human interaction, on surprise.

Our work needed to be made to stand still. We needed to 'record' and 'make imprints' of our experiences, albeit provisionally, such that we might comprehend their meaning and assess their significance. In this desire there is a particular form that our reflexive practices took: a *narrative* one. In short, we craved the kinds of thin spaces in which we might let form happen, and take a pause. This desire to commit to structure, to make textual imprint, to bring temporary closure, is an act of narrative: the performance and concomitant structuring of past exchanges within the present one. Narrative allows us to tame the profusion of exchange. Because of this we crave it (Reid, 1992). As Ben Okri tells us (1997: 113):

> (...) when we have made an experience or a chaos into a story we have transformed it, made sense of it, transmuted experience, domesticated the chaos.

Our feelings of overload told of a need to transmute our experiences from raw text and deep feeling into structured and emotionally secure stories. Remember 'our heads are buzzing and we need to try and make sense of it'. But why should this be significant? Is this just not more evidence of the relationality of methodological practice? Although our own exchanges took relational form, just as the exchange did with

Alison's Mum and Dad, and involved narrative, we think that a distinction can be made between the reflexive practices exhibited in both these situations.

The difference lies in the *awareness* of our own narrative, our conscious reconstruction of exchange into narrative structure, our intention to make textual imprints and articulations of past exchanges. We have been highly trained in the humanities to be aware of our 'writing', of our own 'écriture' (Derrida, 1976) of social reality. We know that we need to construct our knowledge – our relations with the field have made this a necessity. We are aware of our need for the taming of the 'wildness of diversity' (Geertz, 1973). The ontological condition of future writing and analysis was always present in our fieldwork. Were it not for the intention to write we would not have been there, in that precise way, at that precise time. To maintain otherwise would be to impute to the object the manner of looking, a criticism Bourdieu levels at Geertz for his famous, even infamous analysis of the Balinese Cockfight.

This has implications for May's nomination of a referential form of reflexivity. May (1998: 157–158) takes the latter to signify:

> the knowledge which is gained via an encounter with ways of life and ways of viewing the social world that are different from our own. The constitution of this discursive knowledge and its implications for social practice, enables an understanding of the conditions under which action takes place.

Our data enables us to differentiate the conditions under which this referential reflexivity takes form. Referential reflexivity is a fundamentally *differentiated* form of social action, which exhibits, in our data at least, *crucial temporal and spatial contingencies* largely dictated by relations in the field.

We found ourselves back in our 'thin places' at the Red Cuillins, or in our habitual, quiet and spacious café again and again throughout our research, having found time and space to come together in ways that enabled us think and represent, to pause, to rupture the flow of data collection and to write, away from the distractions where thick descriptive potential was writ large at every twist and turn. Whereas the exchanges with Alison's parents involved narrative through exchange, the turns detailed earlier in this section display *the conscious placing of exchange into narrative.* The time and the space had come for this writing. It was a structure-of-feeling, a 'social experience in solution' (Williams, 1977).

Reflexivity as Collaborative Form

Our research has been substantiated in theory and in praxis as a consciously collaborative enterprise. We do not wish to engage in too much navel gazing and indulge ourselves with stories of each other and the ways in which our collaboration bore fruit – as ingratiating as that may be. However, if we draw analogously from the field of anthropology, it is possible to suggest that our research has been born of a kind of intellectual 'kinship' and that as such, our reflexive practices are analogous practices of kinship too.

In connection with kinship analysis, some general findings by Marshall Sahlins (1972) may be pertinent for an inquiry into narrative exchanges such as our own. He established that social 'distances', that is particular degrees of proximity between kin, determine the nature of different exchange relationships. The closer their kinship positions, the less individuals try to maximise their own rewards at each other's expense. Theft, he asserts, from a stranger is less outrageous in this respect than theft from one's sibling. There is some relevance in this analogy to our own relationship. Research collaboration can, at the best of times, be fraught by interpersonal difficulties between researchers. Our own experiences suggest that our intellectual closeness, at least in terms of theoretical and methodological concern, provided a mediated space for openness, honesty and the reciprocal attempt to maximise each other's reward. In our experience, intellectual kinship was and continues to be a good reason for collaboration.

Augé (1995) suggests that supermodernity may allow for a rediscovery of old anthropological concepts in new sets of social and cultural relations; anthropology should not be afraid that its distinctive way of appropriating the world will have nothing to offer to the analysis of contemporary globalised societies. Instead he suggests that notions such as kinship may return to us, if only we continue to ask the same broad questions of the world. And we may indeed find kinship, in equal measure, in the field of tourism, in the field of reflexive ethnography and in intellectual relations.

Crucially, however, we do not suggest that our kinship is reciprocal in the sense that we both derive the same things from our collaboration, or even where we do derive the same things, that we do so in equal measure. Our collaboration is highly partial and itself based in differential exchange. Ian Reid (1992: 4) makes this point. He writes:

Even at a rudimentary level, it takes two to have an exchange, and what is useful to one will seldom be of equal use to the other. Acts of giving and taking hardly ever form a balanced pattern of intersubjective reciprocity. On the contrary as Jean Baudrillard (1975: 75) puts it 'exchange does not operate according to principles of equivalence'.

Our exchanges also operated without Baudrillard's principles of equivalence, a point most notably demonstrated by the extended discussions we had with each other on the Wednesday night, four days after our arrival on Skye. Four days of sharing a car, sharing exchanges, sharing the beating sunshine, sharing hunger, sharing tiredness and sharing the increasing realisation that despite our similar research interests we were different, in some very fundamental respects, led to an elongated conversation about what we thought the differences were between us and how this impacted on our work.

Reflecting on the importance of texts and narratives as a medium for our practices, there is a further crucial aspect to our reflexive praxis which emphasis on the textuality of social relations underplays. Our reflexive practices were not just sets of texts, cultural constructions or mere pieces of fiction. Rather our reflexive practices were tiring, demanding, smelly, at times irritable, in need of food, drink, a rest, some cakes, they involved long walks up big hills, big bags and small talks, long talks and sleepless nights in snoring dorms, emotional fragility and biological instability, blistered heels and eightsome reels and more cakes. Reflexive practice was historical and material, written through the body, its physiological structures and cultural articulations.

Our reflexive praxis, then, was aware of the extensive shortcomings of purely text-based understandings of methodology as a consequence of the materiality of our emergent exchange ethnography. Reflexive praxis for us has a profoundly material form. We felt our data. We felt its overabundance and the impossibility with the violence of representation; we felt the stories that people were telling us. Our methodological texts were sullied by their material form. Collaboration was material and shared. Methodology is a profoundly human activity and crucially dependent upon the human relationships in which it takes form. With all this in mind, let our travels begin – at home as we pack.

Chapter 4
Packing

On the Subject of Packing

Every year, dear, valued reader, when autumn comes and the trees lose their leaves I find myself drawn to Scotland. When you read these lines I will, without a doubt, be in that fascinating country already, together with my VW bus and my luggage. And that is what I want to share with you today – my 'packing orgies'!

I should say in advance that I live in, out of and from my bus. That means that everything that I need must have its place and this is also the case for me with my many needs (eating, sleeping etc). These are demanding starting points as five weeks in Scotland, with mountain climbing, the visiting of much mud and isolated trips to islands etc., demand considerable equipment. Off we go then – but let's not rush it. First lists are written, added to, things are crossed off, added back on, clothing, utensils etc are sorted, piled up and after some shaking of heads, re-stacked.

It carries on like this for several days.

Then, finally, the starting gun: space is made in the bus. This is a hopeless undertaking, what with all the pre-sorted piles of everything conceivable: After all, as my esteemed Mr Murphy, of Murphy's Law, says, when you open a can of worms, you'll need a much bigger one to get them all back into. So everything is sorted again – the fatal problem at this juncture is that whilst I'm sorting I always think of new, indispensable objects that I have to take with me.

At some point, long after reaching total exhaustion, everything has been stowed away. The toilet is on the roof, tools are under the seat etc. Actually finding things I usually just leave to chance once in Scotland. The last possible moment of departure usually comes about one hour before finishing packing (after all there is a ferry to catch in Amsterdam). So I race through Germany like a reincarnated Alberto Ascari, and think of Aunt Jolesch's wise words: 'Departures always happen in a rush.' (http://www.zebra.or.at/zebrat/41artike17.htm)

Packing, as we see in this extract from a web travelogue from Germany, is the subject of stories. The stories that get told on the subject of packing, like those of tourism, are both big and small, regarding the minutest of details right up to subjects of life and death.

In this, and in the two subsequent chapters comprising this section, we seek to explore, theoretically, and through our data, the different material and nonmaterial aspects of culture that get packed. The mood of this section is therefore deliberately anticipatory. This chapter is about all kinds of material aspects of packing and preparation. It is about the journeys that occur before actual departure. It focuses on what we term 'the work of preparation' and the changes in routine that typify the move from the habits of everyday life at home to other routines and modes of being that are part of tourist life. This aspect of 'the work of preparation' is important for our discussions in this book, as it enables us to problematise binaries between work and leisure, a subject we shall return to in Part 3.

Throughout this chapter we draw on a range of literature from disciplines beyond the field of tourism studies, and continue the work begun in the previous chapters of grounding concepts in our under-standings of tourism, both as generic and as particular. When we speak of cultural baggage in this chapter we do so in order to make a distinction between the items that actually, physically end up in the travel bags and other less tangible cultural aspects such as languages, attitudes, education, knowledge, research, reading, stories, traditions etc., that are bagged in the memory, the mind, the experience and the body. All these things have traditionally, though often problemati-cally, been lumped together into the anthropological categories of culture.

We do not wish to suggest that these less tangible aspects of culture do not have some material aspect to them. Spoken languages are of course carried in the body, are physiologically produced and therefore have a material quality. However, for the sake of (provisional) analytic clarity, this distinction between the material and the metaphorical baggage will be used throughout this section. We begin our investigation of packing with some definitions of packing and with the specific and general questions our data and experience raised for us in this context. We then present a selection of the material and cultural items that get packed before turning to some broader cultural discussion and analysis.

Packing

> **Pack** *v.* put (things) together into bundle, box, bag, etc., for transport
> or storing. 2. Put together closely, crowd together. 3. Cover (thing)
> with something pressed tightly round [...] 5. Fill (bag, box, case etc.)
> with clothes etc., load (animal) with pack; cram (space etc) fill
> (theatre) with spectators [...] 9. Depart with one's belongings.
> (Oxford English Dictionary)

In the context of tourism, packing is not just the action of putting things
together into a bundle or box or bag for transporting, nor is it just the
loading of a means of transportation – a bicycle, a car, or someone's
back, with a load to last for the duration of the holiday. Packing, for the
tourist, is not just about making belongings tightly secure, or filling a bag
with belongings before departure. Packing, for tourists, is ontological as
well as epistemological. In other words it is an action that is both
concerned with how we understand our ways of being in the world, and
how we let our knowledges of the world affect our actions. Packing is an
intimate activity and a whole performance about the world as we think
we know and imagine it.

The specifics of materials that get packed make certain actions in
packing common to particular cultures. The particular shapes and
technologies of objects e.g. knowledge many share of a brand of shower
gel that is popular but may leak in transit, leads to specific preventative
measures taken when packing, which other cultural brands may not
require. Other objects, such as T-shirts, are more universal objects, with a
form that maps onto the shapes and sizes of the human body.

Packing is not something that escapes the researcher. We both lost
objects while packing, and both of us at times shared the experience of
packing together, learned new techniques of packing, and conversed on
the subject of packing whilst conducting our research. The ways we
packed and what we chose as the receptacles for our packing opened out
our prior experiences of travel and the experiences of our lives, past and
present. As we commented:

> I got this bag at a conference. It's actually really useful, surprisingly,
> for a freebie.

And:

> I bought this pack as an emergency when the straps on my old one
> broke just before I went hiking for a day. It's been a really good buy.

Our bags triggered stories in the flow of ethnographic action, something we shall consider in more detail in Chapter 5. So did their contents. The moment of horror looking into the boot of the car in a hotel car park and suddenly believing that a set of crucial pills had been left in the hostel produced panic out of all proportion to its importance, but acts as a reminder of a vulnerability that asserts itself when on the road.

Packing for a trip has parallels with packing for fieldwork. Our own bags also contained empty containers; files, folders, boxes, blank floppy disks, new films for the camera, pristine notebooks, pens full of ink and sharpened pencils. All of these got packed into the kinds of bags that we felt would serve the practical purpose of gathering materials and of participant observation. We were also anticipating and imagining a future, different scenarios in which 'having things handy' would be necessary for the professional work of gleaning data from tourist contexts, to be unpacked and repacked later during this very process of writing and analysis.

These observations about packing, together with the opening story about packing orgies, leads us to ask: how is it that the activity of packing has become imbued with such a desire to control, to compact, to bundle away? What does the washing of clothes, sharpening of pencils, purchasing of new things, the listing, sifting, stacking and sorting tell us about preparing to do tourism? More importantly, for our purposes, what does our packing tell us about intercultural exchange?

What Should I Take?

Packing is the subject of considerable advice, found on internet sites and in guidebooks. It is often part of the ritual of packing to phone friends or family to ask for tips, should a similar journey, to a similar place, have been made already. 'What should I take?' is a question born of concern, care, insecurity, uncertainty, all moods that are part and, indeed, parcel, of anticipation prior to departure. Equally, the anxieties that accompany packing are joined by a confidence that someone will know what to take, someone will be able to give advice.

Where such packing advice is present in guidebooks bought and sold on the open market, it is as an anonymous part of the package in the particular book. Internet sites and phone calls to friends are of a different order however, indexing something of the potlatch principle. Potlatch is a shared meal in which everyone invited brings something to the table to make a complete meal. The question of what to take is more interesting to us in such a context, as it works with wider concerns regarding the nature and practice of exchange, particularly the exchange of knowledge. The form in which such packing advice is given, as we see below, is not simply in the form of factual lists of items, although websites and tourist companies organising tours do often give out lists of objects to take on holiday (http://www.traveltoscotland.50megs.com/packing_list.html). The forum of internet exchange, of advice, as with the phone calls to friends and family, demonstrates an alternative economy of knowledge, exchanged freely and interculturally. Let us examine some of the advice given.

The material presented below is taken from the website 'virtual tourist.com' (http://www.virtualtourist.com). It contains a whole section entitled 'Packing Lists'. These extracts present advice for travel to Scotland, organised, by those posting advice, into categories.

Clothing/Shoes/Weather Gear:
Something to wear when it's raining! Because it does – it's a fact of life here in Scotland – but the scenery is just as amazing whether it's raining or sunny!
Watch out for all 4 seasons in one day – a typical occurrence in Scotland – just recently we had 10cm of snow at 9.30am, no snow by lunch time-it had been washed away by all the rain, thunder, a hail storm and sun by the afternoon!

Toiletries & Medical Supplies:
If you are going up north during the summer try to have long sleeves and long trousers to wear – particularly at dusk when the dreaded midges come out! These are very small flies that fly around in huge packs and give you lots of itchy red dots!

Photo Equipment:
Always bring an extra camera on a trip like this. My new digital camera disliked the seawater and bled green stuff before packing up. For the rest of my trip my old Canon did the job. I was very happy I decided to take it along.

Clothing/Shoes/Weather Gear:
Bring an umbrella just in case. I was lucky, all the time I spent on the west coast it only rained twice.
Pack your dancing shoes in case you end up at a ceilidh!

Toiletries & Medical Supplies:
Bring a good insect repellent or those midges will eat you alive! They actually have teeth!

A wife of one of the members of Colin's pipe band used Avon's Skin So Soft spray on the band members, spraying it under their kilts:-), and it seemed to work. I was lucky and didn't get bothered by them, but everyone else was.

Clothing/Shoes/Weather Gear:
Definitely take some very warm clothes and a good rain jacket. We found that umbrellas were useless because the wind blew them inside-out too easily.

Photo Equipment:
Take lots of film because you will use it easily!:)

Luggage And Bags:
In the large cities, pick-pockets are everywhere! Make sure that if you are going to carry a back pack, that your pockets have nothing of value in them. Always be conscious of who is walking behind you and for how long.

Also, the tourist shops are small and it is very difficult to have a large bag and walk around comfortably. A side bag or purse is a must.

For luggage, bring what is comfortable for you. Wheels are always nice, since you might have to walk a way to your hotel.

Clothing/Shoes/Weather Gear:
Even in the summer, Scotland can be cool and wet. Make sure that you bring raingear because you're sure to get wet otherwise.

Tennis shoes are good if you plan to walk in the hills. Otherwise, in the city anything will work.

Toiletries & Medical Supplies:
Make sure that you have a first aid pack with you. Medicine can be very expensive there!

Photo Equipment:
Extra film is a must! Film can be up to $15!! a pack. Please, I beg you, bring your own film!

A different website gives more detailed descriptions and stories in response to requests for advice on what to pack when travelling to Scotland.

Don't let yourself be put off a cycling holiday in Scotland by car-driving softies. Even the time of year is dead right. Twice I was in Ireland in September, and once in Scotland (until into October), both times for about 5 weeks, the weather was always first class. Of course we had rain sometimes, you need to have rainwear with you. At the end of September it can get a bit cool, but it's never too bad. Just watch out that your rainwear isn't like a water-resistant plastic bag because you'll be soaked with sweat in it. You should certainly try out your gear here [in Germany] first! I prefer the Trangia spirit-fuelled stove, it isn't sensitive to the wind, but everyone has their own preferences. Hi there Skyediver [an emailer] you are going without a tent?! Is that because – no tent, no floods. Take a tent, no question, and not one from *Poundstretcher* that cost 8,95! Your sleeping bag needs to be comfortable down to at least 0 degrees (if you want it to be cosy add 5 degrees to the manufacturers' instructions), you also need a thermarest, not so much as a mattress – you'll pretty much always find a comfy bit of heather, but rather as protection against condensation. Wherever you've been lying, or had things on the ground sheet, it will be damp and there is nothing worse than a damp sleeping bag. In case your travel bags aren't completely waterproof put everything in plastic bags! And keep checking them for holes. Have a look at www.rad-forum.de, it has lots of tips, but use the search function before you ask about things you need. Have a great time planning, Oh I'd love to go back again! www.virtualtourist.com

These postings are full of display, advice, practicality, humour and, above all, narrative. They are acts of imagination and acts of experience of material circumstances. They are the work of guests and of hosts, as tourists, together. They display practical knowledges – in action. Some are highly technical, such as the advice above, with exact detail given regarding ways of keeping out the Scottish water when camping or cycling. Tips are followed up rhetorically, however, with stories and emotions. 'Take this kind of plastic bag, because when I did this trip I found that I had this problem with condensation', for instance.

Opening a Space for Fiction

What we see in action here is both the imagined and the actual *emplotment*, in Ricoeur's (1984) sense of an *opening of space for fiction*, around the material advice. In other words, what these exchanges demonstrate is a bridging of the gap – through narrative – between real, lived experience of the world and the sense that has been made of these experiences, now in the form of the communication of advice and stories. The exchanges that occur on message boards, between travellers to Scotland, are exchanges of things of value.

What we see here, then, is the framing of valuable practical knowledge in narrative. The knowledge is wrapped in all kinds of narrative structures in order to serve a variety of purposes. We may characterise these, based on the advice exchanged in our message board extract previously, with the following taxonomy (Box 3) of the purposes of advice. This is not an exhaustive list.

Intercultural communication, when it takes form in tourism and involves the moving of both the tourist's own body and also the artefacts that the tourist chooses to pack for the purpose of being a tourist and doing tourism, involves a prior set of preparation and exchange. These generate their own discursive regimes and narrative forms, as we see in the examples above ('our umbrella blew inside out').

All of the characteristics of the advice given point to a concern for the tourist body and the sense of what we may term spiritual or emotional well-being. They also show an anticipatory concern for the materials that tourists will gather, materially and experientially – photographs in particular – from which they may generate their own narratives and play, confidently, as advisers, tip-givers and story-tellers themselves, in the exchange forum, upon their return.

What is exchanged then, and of value, is a cultural capital, displayed in the form of certain narrative devices and as technical expertise. This is

Box 3 Characteristics of exchanges soliciting and giving advice

1	Display
2	Concern
3	Help
4	Reassurance
5	Memory
6	Pleasure
7	Offers to lend
8	Friendship
9	Health and well-being
10	Practical tips
11	Rhetorical and narrative credibility
12	Control

advice freely given and which demands nothing more than a community of listeners, who solicit advice, or who respond to advice given with their own embellishments. It fulfils the double cravings that Reid (1992) speaks of, for narrative and for exchange. It also genuinely enables the acquisition of practical knowledge regarding what to pack and what not to pack on a trip to Scotland.

Packed Goods

When we come to examine the items that were packed in the baggages we encountered as part of our research – obviously with subtle, idiosyncratic variations, we find general categories as outlined in Box 4 below.

All of these objects point to key characteristics in the work of preparing to be a tourist in Scotland. They are objects that can make a strong claim on us as tourists, they are required or perceived as literally unforgettable. In the process of packing, touching these objects is reassuring, ordering. It stills the anxiety, calms the nerves, and triggers memories of other places where such objects accompanied us, at the same time as enabling us to imagine ourselves with these objects, in new

Box 4 Packed items

Hygiene and beauty products
Money, purse, wallet (often specially designed)
Clothing
Food
Passport/documentation
Gear – technical, specific
Adapters
Guidebooks, maps, information
Phrase book
Reading matter
Recording devices – cameras, camcorders, digital equipment, notebooks
Brochures
Personal trinkets, jewellery

ways, in the future. In short, past and future care collapse into the present, as time shifts its shape to accommodate the objects and the packer's desires.

The idea of preparation suggests the making ready of oneself, of another person, or of a thing. It has a precise goal as an activity and state of being. The aspects of preparation involved in packing, as presented here, are revealing, we would argue, of different dimensions to doing tourism.

Firstly, packed goods have their own genealogies that are linked to advice given, of previous experience, of the way in which we have been taught to pack and to do tourism, often, in the West, from a very early age. The practice of packing is a learned practice and one that, as a live performance, is always slightly different depending on the audience we are imagining ourselves to be encountering, both on the tourist stage, and also once the holiday is over. Examining the objects we take with us tells us about our pasts as packers – are we experienced technically, do we know how to read maps, do we have the language already fluently in our bodies or will we stumble through phrase books, do we know if we can buy everyday items like toothpaste, suncream etc? They also tell us something about our future hopes – do we have audiences in mind, are we intending to write home, send postcards, keep a log to serve our memories in a future age?

Secondly, certain rituals and daily practices of preparation of objects and of ourselves are present in our packed goods. These involve washing, cleaning, making ready, folding for the best possible tourist performance. Kinds of comfortable, domestic relaxation around our clothing or technical gear are not what we find. Instead, as we saw in the earlier description, we check our bags for holes, we sharpen our pencils, we stock up on film. Here we see a projection of desire for the best possible performance, onto the goods, whilst we are away. There is no room for error, we must be ready, prepared, or we may be washed away, sunburned, inappropriately attired, we may have no record of our trip. The objects in our baggages, and the cleansing rituals that accompany both these objects and ourselves point to the fact that we are leaving our own civilised orderings and, however temporarily, creating new mini-civilisations of our own, in the unknown. We use the notion of civilisation carefully here, grounded in the anthropological insight that dirt is disorder (Douglas, 1966: 41); 'if uncleaness is matter out of place, we must approach it through order'; and that rituals of cleanliness are at the heart of the orderings of social life.

Finally we see a further projection of this last aspect here in the form of an anxiety vis-à-vis the unknown. The sense is that the unknown will require a certain vigilance and so the objects we choose will enable us to maintain certain standards, for instance, of personal grooming, of education, of safety and well-being.

These, then, are the items that Brecht refers to, in the context of rehearsal process for the theatre, as the 'least rejected items'. That is to say, these are the kinds of things that through the performance of packing, the listing, sifting, trial runs and rejections, will eventually be chosen as the ones that may best enable the practical performance of being a tourist. They are also the ones that, for our purposes, enable the successful display and exchange of knowledges, recorded and therefore verifiable, upon return. Note, for example, the loss of a camera on the trip described in the packing advices earlier, and the tourist stories of films lost, jammed cameras or film-developing disasters that often form part of the content of narratives of holiday experiences.

The potential disaster here is not so much one of a risk to the life of the body, but to the life and cultural veracity of the narratives that need to be formed upon return. In other words, what is at stake here, reflected in these material objects, is the potential narrative capital, and, paradoxically, a future orientated care for memories not yet experienced, for the future back home, once the holiday is completed.

Benjamin's Books

The working method of this chapter owes a debt to Benjamin's description of the activity of unpacking. The aim, of course, through reflection, theoretical and empirical analysis and recursion, is to unpack the activity of packing. This dialectic is at the heart of this chapter and the structure should be understood as one that begins with the activities of preparation and packing, which build up to departure. These activities include learning, consuming and performing.

Walter Benjamin, in _Unpacking My Library_ (1973: 61) shows us something of the state of mind that accompanies his experience of _un_packing:

> I am unpacking my library. Yes, I am. The books are not yet on the shelves, not yet touched by the mild boredom of order. [...] Every passion borders on the chaotic, but the collector's passion borders on the chaos of memories.

Unpacking, in a new home or upon a return home, is an activity, for Benjamin, which is overwhelmed by the experience of being and memories of being. It is an activity that expects order, but does not yet experience it, and which reveals itself in the chaos of objects and memories, in the random nature of handled objects dug out of the case. It is an activity that is passionate and where those passions are stimulated by the opening of cases, the handling of objects which have been carefully chosen and to which stories accrue:

> Now I am on the last half-emptied case and it is way past midnight. Other thoughts fill me than the ones I am talking about – not thoughts but images, memories. Memories of the cities in which I found so many things: Riga, Naples, Danzig, Moscow, Florence, Basel, Paris: memories of Rosenthal's sumptuous rooms in Munich [...] memories of the rooms where these books have been housed, of my student's den in Munich, of my room in Bern, of the solitude of Iseltwald. (Benjamin, 1973: 68)

We are not given an insight here into Benjamin's experience of packing but we do begin to understand something of the power that possessions hold, and the added power that comes to them as they are packed and then unpacked, as they are made chaotic by movement and create a chaos of memories. This mood around unpacking is, to Benjamin, a delighted, almost elegiac fascination with the extraction of objects from their old contexts:

> Nothing highlights the fascination of unpacking more clearly than the difficulty of stopping this activity. I had started at noon, and it was midnight before I had worked my way to the last cases. (Benjamin, 1973: 67)

Packing, as opposed to unpacking, is the reverse of this experience. It is about temporarily removing possessions from their home, it is not an activity that is savoured, where newly found objects are lovingly turned over and admired amidst the flood of memories from their acquisition on a journey. Packing is less leisured. Each of the objects selected may yet acquire an extra layer of fetish, may yet become a site for narrative, but as the stories are not yet known, so the activity does not allow the chaos of memory and the disorder of unpacking. Instead it controls the chaos with the 'boredom of order'.

This chapter has examined material aspects of the activity of packing, as both a structuring device in preparation for travel and as actual instances from our empirical work. It has looked at the ways in which we pack and the technologies and genealogies of packing. It has looked at where we gain some of our knowledge about what to.pack and how to do so, from message boards to mothers. It has also explored the way that the emotions associated with such knowledge, or perceived lack of knowledge, draw us into exchange. It has examined the materiality of the activity and also the ontological link to the work of preparation, as knowledge is made existential and is used to prepare and to protect.

In addition we have shown how packing, although a present tense activity, actually works both in the future tense, imagining the destinations, anticipating events, and in the past tense, based on prior knowledges and triggering memories. Packing, under our analysis, links us in to ideas regarding structure-of-feeling (Williams, 1977) and emergent tourist tenses and grammars, to which we shall return in subsequent chapters.

Packing, we have argued, is worthy of scrutiny because it instantiates, materially and imaginatively, the work of preparation that precedes our holidays. We speak, for example, of 'living out of a suitcase' when on holiday, so the work of packing prepares us for this highly circumscribed material mode of being. For our purposes, more precisely, an examination of packing allows us to consider how it is that the items we pack and the ways in which we pack them already demonstrate aspects of intercultural exchange in action. In other words, the contents of our suitcases tell us stories about our expectations of and readiness for encounters with other cultures and other languages, as tourists.

This consideration of packing also enables us to reflect on how it is that we learn to imagine destinations, eventualities and how the items we choose to put into our suitcases reflect the things, at different stages of our lives, we consider to be most necessary. The items, for example, that children select for journeys, of their own volition, do not resemble those chosen by their carers. The favourite teddy bear, however, although not materially necessary for bodily survival, may wreak untold emotional havoc for a holiday if forgotten, or, even worse, lost on a journey. Stories told of holidays where teddy bears were lost often, sadly, become a most memorable aspects of that tourist experience. It is to these aspects of what happens to the stuff we pack that we now turn.

Chapter 5
Packers of Culture

It is another glorious day. All our guidebooks suggest that a five-star activity is a 10-mile walk on the peninsula, to the North. We've been staying at the Flora Bay and notice a certain degree of incongruity between our hiking gear, a couple of water bottles and a worn rucksack, and the smart touring cases of our fellow guests. We feel a little embarrassed or ashamed, even out of place and uncomfortable around some of our luggage.

Once out of doors and covering ground there isn't a problem. Our bags work, we can carry all our gear with ease and appropriateness. Everyone we pass looks like us. The day is data-rich and full of intercultural encounter. The camera fills up with images, taken out and packed away repeatedly. Every now and then we stop and take out our mini-tape-recorder to record observations. More usually however, we talk, watch and scribble notes. Around us we see displayed, on bodies, the results of packing and unpacking. T-shirts with text emblazoned, varieties of light hiking boots, cameras, maps, binoculars, walking sticks, lots of different daysacks. Suncream is applied from bottles bearing a variety of European languages.

At one point we squeeze between sharp rocks and one of our packs gets scraped. The marks linger on, physical reminders on every trip of a walk in the sunshine, on Skye.

At the end of the day we are dusty. So are our packs. We are sweaty. So are our packs. We are travel-stained and in need of refreshment. We've drunk all our water, our packs are lighter. We've eaten our packed lunches and know we will need to restock and pack our food all over again tomorrow. Our cameras and binoculars have spent the day in and out of their cases. We've finished off one bottle of sunscreen and empty snack bar wrappers are stuffed into side pockets, ready for the litter bin at the car park.

Sitting on the Suitcase

In our previous chapter we examined some of the ways in which packing enables us to reflect critically on the work of preparation that precedes our holidays. It is not just the packing of material objects that show something of the multifaceted material and imaginary worlds of intercultural exchange. Packing is a physical activity involving grunts and groans and material goods. Once we take our bags on holiday they become travel bags, objects made to move, with their contents, in new, unpredictable cultural contexts. Both bags and their contents are changed by this movement and they are consequently made available for new meanings.

Packed objects have a symbolic status. This status is not fixed, but constantly shifting, throughout the spatiotemporal experience of a holiday. Many meanings accrue to those items finally selected for the road. As these items take their part, or not, performing their allotted task, or not, during the course of a holiday, they assert an agency, semiotically, sensually, politically. Packed material, the agency and the symbolic meanings and interpretations that accrue to stuff, in the tourist context, is the focus of this chapter. The aim is to approach baggage and bags as phenomena, examining how it is that they work for us, do tourism on our behalf and participate in intercultural exchanges.

Obviously, all material items can become cultural, but there is a sense in which, metaphorically, we also 'pack culture' when we set off on holiday. The chaotic sifting, sorting and choosing that is applied to materials may also work analogously for thinking about cultural baggage. Does 'culture', however fraught and contingent the term, end up folded, stuffed and squeezed? Is it also squashed down when we sit on it to make it portable and manageable for journeys? In Chapters 6 and 8 we examine these questions in the specific context of travel guides, their usage and their work in doing tourism. In Chapter 6, our focus is on the written narratives and guides of tourism, on tourism's textuality. In this chapter, following a discussion of the hermeneutic aspects of material baggage, we look specifically at oral narratives and the meanings made by us when we take stories out of our bags and our bodies. In other words, we work with certain moments from our fieldwork that triggered feelings, memories and stories as a result of encounters with bags and their contents.

Objects, in these oral moments of tourism, take their place as characters and catalysts for wider stories. These oral stories, phenomena in themselves, like the material objects that underlie them, lead to further narrative exchange. The aim, then, is to present these phenomena and work with these reflexively. In this we are conscious that when travelling with culture, orally and materially, we are not travelling with an abstracted, reified concept. This, as Abram (1997) makes clear, comes as oral stories are written down, dislocating stories from direct perception and direct material resonance. As tourists we are mobile story-tellers, in touch with new places but not yet consigning these places to written text, and thus bearing their imprint. The written stories of tourism come into view later in Chapter 6.

Just as packing has material and metaphorical resonance, so too does the travel bag. It is not only what goes into the bags that is of interest, but the bags themselves. Technologies of portability have accompanied the growth in travel and tourism. The activity of packing is accessorised and the symbolic nature of the containers is worth some reflection. Old bags, new bags, rucksacks, suitcases, bags that roll, bags that sling, bags that fit car boots, lockers, compartments, bags on backs, shoulders, bums and belts – all have their stories to tell and all are props in the packing performance.

And what of the packer of the bags? Our possessions, chosen for the holiday, are also our responsibility and revealing of our status as 'responsible' tourists, citizens of the world. This is the work of choice, highly controlled and controlling – or at least attempting to be. It is work of great importance. If something gets_forgotten, as the anxiety of the packing situation suggests, then it could, of course, be dangerous or at the very least mildly inconvenient. Packing for heterosexual couples and for families has also traditionally been highly gendered work, the work of the wife or the mother to whom the responsibility of preparation falls.

This work has three aspects to it, two of which we explored in Chapter 4: (i) the responsibility for the meeting of needs on holiday; (ii) the responsibility for the narratives and the recording for memory once back home, but also (iii) as we shall see in this chapter, for what Williams terms the 'structuring of feeling' (Williams, 1977). In other words, for the continuing reflexive, social work of making sense of materials that have been taken out of their domestic context and turned into baggage. Baggage, after all, only comes into existence when we travel. It is not something we have, by definition, during nontourist life, at home.

Packing, baggage and its interpretation reflects a certain structure-of-feeling, a set of social experiences in solution:

> For structures of feeling can be defined as social experiences in solution as distinct from other social semantic formations which have been precipitated and are more evidently and more immediately available. [...] and it is primarily to emergent formations [...] that the structure of feeling, as solution, relates. Yet this specific solution is

never mere flux. It is a structured formation which, because it is at the very edge of semantic availability, has many of the characteristics of a pre-formation, until specific articulations new semantic figures are discovered in material practice: often, as it happens, in relatively isolated ways. (Williams, 1977: 133)

Seeing Stuff, Reading Stuff

Travel Bag 'small bag, carried in hand for traveller's requisites' (OED)

The definition in the OED uses the term requisites: '*n*. required by circumstances, necessary to success'. In the previous chapter we explored the way in which packing is part of the imaginative preparation of establishing what are the 'least-rejected objects' for the performance of tourism. Requirements, necessities, success – all are present as hopeful feelings and expectations in the imaginative process of packing the travel bag.

Heavy rucksacks were just one material feature of life on a budget tour on Skye. We carried them, we gave lifts to hitch-hikers with them, and we watched others carry them too. We experienced a wide variety of travel bags, in many different ways and in many states of readiness. Let us examine just three of these.

Three Scenes

(1)   The car, a recently washed blue Peugeot 306, diesel, is parked outside a Glasgow tenement flat. Inside the boot is a medium-sized, navy blue, zip hold-all, a black, synthetic laptop case, a plastic bag with crisps, fizzy drinks and chocolate, a smart leather briefcase, still looking new, and a stout pair of worker's boots. These items, belonging to our first packer, are arranged in an orderly fashion on the left-hand side. There is still plenty of room. To these are now added a second set of bags, brought down from the second floor apartment, in one go, by the second of our packers: a 55-litre purple rucksack, adjusted to 35-litre capacity and designed for the female frame; a small, well worn daysack, blue and black, another black, synthetic laptop case, a pair of well used hiking boots, a small plastic bag containing a sturdy water bottle, some trail mix, cereal snack bars and two bananas and a money belt.

(2) The bicycles, lined up outside a Visitor Centre, Skye. On their frames are specially designed panniers, for back and front wheels. On their handle bars are small pouches and map cases. All the panniers are weatherproof. Tents and bedding roles are stacked on bicycle racks.

(3) Flora Bay Hotel. 10 a.m. Luggage on wheels, leather designer travel bags, carpet bags, suit protectors, golf club caddies are stacked in small piles around reception. A porter is on hand to help owners take their luggage to their cars – Mercedes, Audi, saloon cars, 4 × 4s.

Travel bags, like all accessories and commodity forms, lend themselves to social and cultural interpretation and classification. Their linking to particular people and the ordinary, everyday activities of their packing and unpacking make them objects of social and cultural significance, open to a variety of interpretations and analyses. These classifications, however fluid and provisional, are both those of careful analysis and those of instant processing. In other words, they are subject both to practical knowledge-in-action and to imaginative knowledge-in-action. We may begin, in the tradition of Geertzian anthropology, by understanding travel bags as 'texts', made available to varieties of human beings caught up in tourism in varying ways, for reading.

Loaded with luggage: Touring bicycles outside an island cafe

Barthes (1970, 1975) helps us dividè texts between the 'lisible' and the 'scriptible' (readerly and writerly), in *S/Z*, in order to avoid hierarchies of fiction, and, in order to move the focus of production of meaning from the author to the reader. Understanding the tourist gaze as a semiotic process has proved a powerful starting point for analysis (Urry, 1990). Tourists are semioticians, ordering and producing and affecting their experiences in historical, economic, social, cultural and visual ways. Following Foucault, as noted earlier, Urry analyses the particular, systematic ways of 'seeing' that accompany tourism.

Under this analysis, travel bags get 'seen' by tourists in different ways. To return to our examples, we have bags 'seen' as posh or practical, as conforming to different needs that different bodily activities associated with forms of tourism demand. Those bags at the Flora Bay were instantly read as conforming to the demands of a luxury country house hotel and as working socially and practically to fulfil these needs. Their form works with the demands of car boots and casters. Bicycle panniers tell a different story, of weatherproofing, weight bearing and technical design. 'That's really neat', we may say. Bags are 'seen' as fit for certain purposes, and when bags are seen on shelves or in bags themselves, they are selected or rejected, as commodities, to accord with the demands of practical knowledge and social status. There are certain bags that we, as tourists, would not be 'seen' dead with, and others that accompany us on every trip, for years.

There are limits too, to baggage. The limits come from within and without. Some of the limits are simply physical. 'I cannot physically carry any more stuff'; 'I can't get this jacket in to my case'; 'I'll just have to leave this book behind, I probably won't read it anyway'. They come when we simply have no strength or technologies left for managing the size of the load. They are eased by the various travel technologies of cases on wheels, travel accessories designed to reduce material artefacts to fit our loads. And limits are imposed externally by those we employ to carry our stuff for us; most notably by airlines, and by the legal restrictions imposed now on all travellers, on what they may or may not carry. Packers have to take cognisance of all of these limits and the new knowledges about the limits to packing. How we interpret these multiple possibilities that surround packing, becomes a pressing question.

Readerly readings of bags would lead us to oversee them, or, perhaps more specifically, to only see them. Writerly readings enable us to allocate bags to gender and class and nation, cultural practice and the everyday orderings of life that act to work with and against the grain of technologies of control (de Certeau, 1984). There are, however, numerous problems with understanding culture as text and of only applying semiotic analysis to 'texts' such as travel bags or activities such as tourism, not least of which, according to Bourdieu's critique of Geertz (Bourdieu, 2000: 53), is the problem of 'imputing to the object, the manner of looking'. By insisting on 'reading' as text, privileging the visual and semiotic analysis, we skew our line of questioning towards the traditional Western biases of perception, hermeneutics, texts, and to the dominant practices of humanities scholarship. In the process, we ignore oral dimensions:

> Since stories get told in a great variety of situations, the mere notion of textual exchange may be too pat, too banal, to explain what goes on across that range of diverse socio-historical circumstances. (Reid, 1992: 3)

There is a certain hierarchy of values assumed in making distinctions that are seen as readerly or writerly. Those doing the writerly work are those who do the 'real' and 'important' work. They are the true scholars of cultural texts. There is nothing immediately wrong with this in terms of a way of approaching a question, but only if handled reflexively and in the context of other human faculties that may also be brought to bear. Touch, for instance, or smell.

The Ordering of Stuff Orders the Packer

Packing is not only an activity in which tourists assert agency over objects. It is also an activity that orders packers. It is a process that is changed by material objects along the way. Travel bags show all the signs of this social and material work. Our encounters with the work of packing lead to narrative just as narrative leads to packing. The material manifestation of 'being out of place' at the Flora Bay Hotel, was a well worn, trusty rucksack. It ordered us as packers with certain experiences and expectations. It was one of the things that felt exposing, as though our cover would be blown to other guests and staff, as fraudulent tourists, lacking in golf clubs, leather and leatherette. What we had with us in this social context was sufficient for the outdoor tasks of ethnography, and for the youth hostels and bed and breakfasts, but it exposed us as failed social packers here.

This kind of ordering was, we discovered, a constant hermeneutic feature of tourist life on Skye. Our tourists were constantly processing the bodies and objects of other tourists, including ourselves, to make their own sense of who we were and to add their own imaginings to the stuff that accompanied us. Alison's wearing of a German university T-shirt in Bluewater Youth Hostel ordered her as someone with links to Germany. It also spoke of her education and her use of a fluent German, marked by living in the south. It classified her into a certain region, a catholic, conservative, affluent region of Germany, beautiful, rural and for many of the other German tourists from the North or the East, culturally distant, unwelcome and intrusive even, in this Scottish place. It triggered questions and emotions in people in different ways.

> (Alison) One of the women in my dorm in the Bluewater hostel speaks German, but with an accent. She is tanned, like her German companion. When we first meet, in the dorm, she speaks to me in halting English. I clock her, before I hear her German, as a potential German tourist. I'm holding a copy of *Baedeker Schottland* when I come down to the kitchen and this time she smiles and engages me in conversation, in German because of the book. I ask her where she is from, and discover she is Hungarian. I'm ready to tell her more but already we are discussing the weather, the bigger story of sunshine in place of expected rain, that is happening outside.

A further example of this ordering activity may be seen in the written extract below, which demonstrates the way in which we pack and what we pack this in, provides us with sets of materials out of which we may begin to construct plots and narratives.

Not only does the activity of packing order the packer, making the packer a character in a potential, imagined social drama for all who attempt to interpret the activity and the packed object. The packed materials also form constituent elements for the emplotment, in Ricoeur's terms, of written narratives of human action.

The extract, translated from a tourist weblog, exemplifies this point nicely:

> It's a great feeling, having a bed for the night at last. I am also really happy that I'm in a youth hostel again. It isn't just, as I've said before, because of the money (£4 rather than £15 per night), but because of the company that I keep being drawn back into youth hostels. You don't sit on your own in a hotel room or try to make contact in some bar with some stupid small talk, but rather you find that conversations just happen spontaneously, about where you are from and where you are going, about the weather, food, the problems you've encountered or that you have in front of you.

Although, if I look more closely at this last point I have to admit to being somewhat disappointed on this trip. On the last tour through Ireland I got to know lots of people, who were cycling, like myself, and who will remain with me in my memories to the day I die. This time it has been a bit different, particularly with respect to my fellow cyclists. Up until now I've really only met two different types of cycle-tourists (I don't mean here the 'normal' ones like myself). The first type are the muesli-freaks, with their fully laden rusty trusty steeds acting as a visible sign of their abstinence from all forms of consumption, and everything else imaginable. They ride slowly by, incapable of returning a smile or nod from their long haired heads and goaty bearded mouths, they are searching vacantly into the distance through their obligatory nickel glasses for some imaginary point on the horizon.

The second category is that of the racing biker, shooting past like a cannon ball, without greeting, on light titanium models, and with no luggage at all, often with one or more friends, and in racing gear. He has a sympathetic smile for the likes of us, presumably (though you can't actually see it). I haven't met the category I belong to this summer. This is the category of those who do see some degree of fun in this kind of cycle tour, as well as something physically demanding: A welcome change from the bog standard holiday, a mix of muscle ache, nature and a long lost scouting romanticism. But the youth hostels don't just have cyclists in them and amongst those others who are there you keep finding good folk to talk to.

(http://www.ernstfam.de/scotla32.htm)

This example of written narrative differs from oral stories as it is less distracted by place and material object and more focused on what Abram sees as the more homogenous and placeless. 'Writing down stories', he argues, 'renders them separable, for the first time.' (Abram, 1997: 183). Both oral and written forms of narrative operate as spatial practices, but they attend to the everyday and to place in different ways. Oral narratives are embedded in the places where they are told, they unfold as present tense theatre, they can be easily interrupted, they are turned around and changed easily by material circumstance. Written narratives, on the other hand, enjoy a spatial detachment and an uninterruptability.

Time and Space for Stories

De Certeau, like Ricoeur, maintains that narrative is itself a spatial practice:

> Every story is a travel story – a spatial practice. For this reason, spatial practices concern everyday tactics, are part of them [. . .] These narrated adventures, simultaneously producing geographies of actions and drifting into the commonplaces of an order, do not merely constitute a 'supplement' to pedestrian enunciations and rhetoric. They are not satisfied with displacing the latter and transposing them into the field of language. In reality they organize walks, they make the journey before or during the time the feet perform it. (de Certeau, 1984: 115–116)

Emplotment, to repeat our earlier point, according to Ricoeur, 'opens the space for fiction' (Ricoeur, 1984, in: Simms, 2003: 45), enabling the completion of the hermeneutic circle. In other words, bridging the gap, through time, of our understanding of the world as it is. Bringing an understanding of the world to narrative takes in the world of a 'text' and our world as 'readers', enabling us to order the two together, gathering their fragments into a new sense.

But as we argued earlier in this chapter, narrative is not merely a textual practice, anymore than hermeneutic 'readings' can give us all-embracing stories of tourism. Oral narrative, as we discovered again and again through our ethnographic work, was triggered by material objects coming into the field of the senses, often by coming into view, but sometimes through touch or smell or sound or taste. It was a spatial practice, opening a space for fiction, for the work of meaning making, for the activity of jumping to quick conclusions about people, their stuff and their biographies, through narratives. It was part of doing tourism.

So just as packing orders packers, so the narratives that spring into life, from the materials that are packed, give insight into the worlds that cluster around those objects. They order the packers and the stuff that is packed, to fit it into new schema and to emplot the new worlds that are revealed by the objects. They enable a civilising of the disorder that newness creates.

Narrative also takes time. Lingering over 'texts' is, as Brueggemann (2000) points out, the work of time, detail, patience and thus of scholarship. But sometimes, again, according to Brueggemann, 'words explode' making themselves felt and available for instant interpretation. Similarly Fabian (1983) discusses the material role of time in the construction of the anthropological object. Lingering, as we shall see, is a helpful term, for our purposes, and complements 'reading' or 'seeing'.

Linger *v.*: To stay behind, tarry, loiter on one's way; to stay on or hang about in a place beyond the proper or usual time, esp. from reluctance to leave.it.
To dwell upon, give protracted consideration to, be reluctant to quit (a subject). (OED)

'Lingering' is a tourist activity. Our tourists lingered over their meals and their stories in the youth hostels, or at dinner and at breakfast. 'We always have cooked breakfast in youth hostels', said one family, 'but never at home'. Up on the peninsula we found tourists lingering over views, stopping to engage strangers in conversation, patiently listening to halting English, taking time to smell flowers, paint pictures. Down in the towns there was lingering in bars and tea shops, in gift shops and in front of shop windows. The flâneur that Benjamin writes of in his Arcades Project (1982, 1999), in the context of 19th century Paris, is alive and well when doing tourism.

Of course this is not usually our perception of other tourists. Tourists are also the dupes in 'tourist traps' – in containers made for tourists. They are the ones 'doing' whole cities, whistle-stop tours, carefully regulated visits that are timed and scripted to the minute, as we shall see in Part 3. These aspects of tourist life are given rigorous treatment in the tourism literature. The truth of the matter is that to be a tourist and do the work that any particular experience of doing tourism requires, means disrupting the spatiotemporal flow of our habitual contexts. For the groups that we were concerned with, time and space were often given over to stories, for reflecting on new experiences, emplotting these, closing the hermeneutic circle and doing the tourist work of making sense of intercultural life. We see this exemplified in the reflections of our German cyclist above.

Languages, Translation, Oral Narrative

Perhaps the aspect of cultural knowledge that gets packed away most significantly in tourism is that of other languages. Languages, when they move into other linguistic spaces, re-emerge as translation. For the most part other languages are not displayed in the habits of our largely monolingual lives in the West. But when we change places and countries, our knowledge of other languages and our prior formal or domestic education in these languages gets called out in ways that do not always feel safe, easy or comfortable. The way oral stories are accented by languages is very different to the cleanliness of an edited text and a written translation. Our own languages, as embodied 'baggage' now, may become much less predictably received, frighteningly so perhaps and marking us as foreign, as 'away from home'. Even those travelling to cultures which speak a variant of their own language, or even within their own culture, will experience something of the dissonance of language difference.

Translation becomes, therefore, a key aspect of the tourist experience. Although Venuti (1995) has argued that the translator is a largely invisible figure, and we see this as true for literature and also for translated guidebooks, phrase books and other linguistic props for tourists, it is not true for translation as an oral tourist practice. Translation is a constant aspect of doing tourism for our tourists in Scotland. It requires tourists to move in and out of languages with varying ease and it requires patience on the part of others:

> Translation [...] can be seen as a form of triangulation or, to use Gillian Rose's terms, a 'broken middle' that prevents a violent and dogmatic synthesis of binary opposites. Staying with translation as triangulation, it is possible to see translation as lying between the pathology of universalism and the pathology of difference. (Cronin, 2000: 89)

The nirvana of intercultural communication, where all cultural differences are erased into a universalist reconciliation of oppositions, that Cronin critiques, is not one we find in youth hostel kitchens, bed and breakfast accommodation, or even, as we discuss later, in the most serviced-scaped of tourist amenities. It is also a charitable work (Williams, 2000) done by others and for others in order, not so much to ease the flow of global capital from tourist pockets and into destination coffers, but to exercise a certain hospitability. It is a way of staying in the 'broken middle' and working with the vulnerable difficulties of halting languages as translation, in tourism.

What is significant, in the context of a discussion of oral narrative, is the creation of tourism interlanguages (Davies et al., 1984) and the negotiation, in language, of the 'language shock' (Agar, 1994) that occurs as our own and other languages are thrown into new relief by doing tourism in particular places. Many of the uses of language we discovered in our tourists, but also in ourselves, led to a mixing of elements of different languages 'So ein Mug möcht' ich auch mal haben' (I'd like to have a mug like that). Or 'Es gibt kein Litter' (there isn't a litter bin here). The oral narratives encountered then would pick up elements of the language of the destination and mix it in with their own as a way of responding to the material life they encountered. Intercultural communication, for our tourists, was not a matter of performing perfectly in English, or even Gaelic, nor was it a matter of refusing to leave the medium of the mother tongue. The differences between languages and cultural artefacts were not masked out, rather they were both foreignised and domesticated at one and the same time (Venuti, 1995).

Stories are told in *languages*. Oral narratives are constituted linguistically and materially in *languaging bodies*. Excited narrators will move their teeth, tongues and spittle in ways that match their imaginations and memories. Language, as Archer determines aptly, possesses the quality of 'aboutness', evoking reality (Archer, 2000: 154–155). Archer argues, through *Being Human: The Problem of Agency* for the primacy of practice as the lever or fulcrum for discursive and embodied knowledges of the world. She also argues that our embodied and our practical knowledge develop in direct interplay with nature and material culture, respectively, and that this entails the many things that humans know have not been filtered through the meanings belonging to the discursive order. This is the work we see our tourists doing, and we find ourselves in our own recorded dialogues. It is the embodied, oral, spatiotemporal work of creating discursive and embodied meanings out of our unfolding practical knowledges of the materials we encounter and that are part of the practices of tourist life. It is about *know-how* not *knowledge about*.

Stories are also told from myriad different positions in the world. Ingold argues that:

> [. . .] learning to perceive is a matter not of acquiring conventional schemata for ordering sensory data, but of learning to attend to the world in certain ways through involvement with others in everyday contexts of practical action. This, too, is how the anthropologist learns to perceive the field [. . .] through becoming immersed in joint action with [. . .] fellow practitioners in a shared environment. (Ingold, 1993: 222–223)

The objects over which tourists linger are the ones that assert their own claims on us in ways that disrupt the usual orderings of habitual, nontourist life. Where Ingold's work helps our project most particularly is in his insistence that human cultural intelligence does not only operate at the level of subjectivity, through the engagement of minds in language, but rather that:

> [. . .] the self is constituted as a centre of agency and awareness in the process of its active engagement within an environment. Feeling,

remembering, intending and speaking are all aspects of that engagement, and through it the self continually comes into being. (Ingold,
1993: 103)

This interagentivity, as Ingold (2000) terms it, is, we may argue, at the
heart of oral tourist narratives. When we travel with culture we
communicate interculturally, in our exchanges with those speaking other
languages, with those of a different social class or of different nationality
or ethnicity. We perceive and read others, their sounds, their objects. But
we do so in a material context that also works its way into our narrative
orderings as part of our active process of engagement with an environment, pulling in the weather, the scenery, the litter, the souvenirs as ways
of expressing emotion and giving hue and shade to our journeys. This is
where we see the interagentic potential in oral narrative as a form of
reciprocal engagement with a range of phenomena.

Conclusion

In its theoretical analysis, this chapter has attempted to chart a course
between the understanding of culture as text and the understandings of
material and oral culture. Drawing on the work of Ricoeur and de
Certeau we demonstrate the presence of place over space in oral
narrative, but we do so in the context of ethnographic and narrative
data that relates to the generative qualities of materials in producing
narrative.

Culturally too, we have argued that we find our dispositions and
knowledges and practices to change character when moved from the
places where they habitually stay. In the everyday rhythms of our lives,
the knowledges and habits we make manifest remain relatively static,
following certain predictable scripts and habits of movement. We know
where to buy our bread, park our car, we know what our home town was
famous for in the past, who or what to wheel out when we have visitors
to demonstrate that our home too has a history that connects it to wider
movements in art or music or architecture, to industry and the stuff of
events. We pack these knowledges away when we travel, not knowing
how they may be called out.

In the big stories that get told about tourism we see models of tourist
behaviour, perceptions of tourists and a supposed response to tourist
needs by the tourism industries – en-suite facilities, service stations and
single serving sugar tubes become the order of the day, the secret of
success, in monetary terms. Interpreting travel bags and their contents is
not so much a way of reading culture as text but of understanding

humans as story-telling packers of culture and anticipators of experience, working with new spatiotemporal relations in new contexts as they holiday, and taking the time and space for narrative.

The new oral narratives that are generated do not exist in isolation, even if their development as narrative form may reflect relatively isolated creative practice. Such narratives are also triangulated, tested out against other pre-existing forms that have shaped the way a destination may be imagined. The stock of cultural artefacts and material is one set of such resources for a tourist's oral narratives, another is the written travel guide, to which we now turn.

Chapter 6
Bag-sized Stories

In the previous chapter we examined aspects of material life that give rise to narrative. We looked at travel bags, both as receptacles for material objects and as material objects themselves. We then described how travel bags trigger a range of opportunities for and taxonomies of (narrative) exchange, circulating widely. We argued for an understanding of packing and of travel bags as part of the practice of everyday life, drawing on both de Certeau and Ricoeur in our attempts to see how narratives of packing and of bags order packers. These narratives are in and of themselves careful emplotments, we argued, helping meaning to be made and creating spatial transformation. Travel bags, we saw, were not only the triggers for narratives and consequently for cultural change, they also were changed by their own travel biographies.

In this chapter we turn our attention from oral narratives to what may be understood as 'written culture' in the context of tourism in Scotland. As we argued in Chapters 4 and 5, packing may be articulated metaphorically. Cultural baggage, as well as material baggage, is taken with us when we travel, both consciously and subconsciously acquired and carried. Continuing our concern with narrative, both oral and written, we examine the big stories, the little stories and the gradations of different experiences of culture in between that accompany tourists on their trips to Scotland. In particular we consider the portable, practical and representational example of travel guides, discussed admirably as historical artefacts by Koshar (2000), in the context of German travel guides. We focus here on their anticipatory function; how these guides are used in preparing for holidays and preparing the imagination of destinations; how they are used in the nontourist everyday.

We focus here, in part, on the phenomenon of German travel guides and we do so for two reasons. Firstly, to build on the historical analysis already undertaken by Koshar by examining German travel guide usage in the present. Secondly, because as a national and linguistic grouping, the phenomenon of German tourists exerted a particularly strong claim on us during our research. Not only were we both able to move easily in German and to draw on our own prior experiences of living in German cultures, but the strongest interlingual dimension of our fieldwork was conducted in and through German. We had many opportunities to use our German and to connect with Germans in conversation, in their language and through their artefacts. In other words, to come back to the point we have made throughout this book, we are responding in this chapter to the particular claim made on us by the phenomenon of German tourists on holiday in Scotland, and to their written narratives.

Apodemic Literature

Travel guides are written and consumed with the intention of freeing the modern subject for travel. They have emancipatory potential and are framed as such. They are also both didactic and instructional and as such may be understood as 'apodemic literature' – the literature of what Stagl (1980) terms *'Reisekunst'* or the art of travel. Stagl defines apodemic literature as works in which the central concern is providing systematic rules useful for travel and observation. In terms of cultural construction, apodemic literature is a literature that should not be underestimated, as it is materially as well as affectively *performative*.

The category of travel literature and travel guides has become a focus of debate in the fields of literary studies, translation studies and cultural studies, in recent years (Buzard, 1993; Cronin, 2000; Koshar, 2000). Travel writing has become an important genre in its own right and has freed itself from understandings which see it as in some way subcanonical and therefore not worthy of serious hermeneutic work. As a result, travel writing has been demonstrated to be more than capable of bearing the weight of critical, hermeneutic scrutiny (Robinson & Andersen, 2002). In the past, travel literature was believed to be a debased form of more populist culture. In more recent times, however, the deconstruction of cultural categories and the sharp rise of the activity of travel in the West have altered this perception. Similarly with the travel guide.

As a paradigm of travel writing, tourist guides have been subject to divergent cultural interpretations. Accordingly, travel guides might be viewed as aides for those who wish for some distinction from package tourists, perhaps seeking autonomy in their travel, but who, nonetheless, also require their travel to be prepackaged, for the sake of comfort and safety, as a preparatory text for the yet-to-come. Although any categorisation of tourists and travellers risks generalisation, it has, nonetheless, been a point of regular note in the literature on tourism. 'Travel' is active, 'tourism' is passive; 'travellers' are 'good', 'tourists' are 'bad'. Although attempts have been made in the tourism literature (Crick, 1996) to deconstruct the problems with such distinctions, they should nonetheless still be taken seriously as they continue to index, in language, ideological struggles over the purpose and meaning of travel itself (Alneng, 2002).

Apodemic literature is a literature which exerts a significant *performative* role, in the Austinian sense (Austin, 1975), upon the reader. Indeed it is a literature which is written and consumed with the precise intention – on both parts – of affecting behaviour. Travel guides have a clear agentic effect on patterns of tourism and tourist behaviour, dictating itineraries and 'tutoring' the tourist gaze (McGregor, 2000). Travel guides, as Koshar (2000) demonstrates, are consequently sites in which struggles over ideology and cultural change occur, indexing competing interpretations of modern culture either as a force for rationalisation and institutionalisation, or for emancipation. Competing interpretations of the role of guidebooks index shifting cultural understandings of travel.

Valentine Cunningham discusses the heuristic aspects of the reading of novels in establishing cultures of travel of varied kinds.

> Novels are [...] heuristic systems, that is places of learning, of self-education, offering positive results to their studious participants. The protagonists in most traditional novels, as in most traditional narratives, are breezy with confidence that something positive is to be gained by their time in the narration. They're on a kind of voyage of discovery simply by dint of their being in the fiction. It's a voyage that has many convergent aspects and analogues the epic journey of adventure and heroism, the colonizing journey of capture and discovery, the Dantesque moral journey of personal discovery and salvation, the journey that sorts out puzzles for oneself or for others. The traveler on such a journey may be variously pilgrim, soldier, picaro, agent of imperialist expansion, detective Ulysses, Quixote, Crusoe, Theseus, Sherlock Homes, Everyman. But whatever particular label on the luggage, everyone involved in the ancient business of story is on a learning road. Authors and narrators the text-producers; characters the subject of texts; and readers the consumers of texts: all of them traditionally travel the same path, in turn, from greater to lesser ignorance, from less to more knowledge. (Cunningham, 1994: 228)

What travel literature and travel guides both have in common is that they are written artefacts of culture and, as mentioned in the previous chapter, they are therefore separable from place in ways that oral culture is not. In addition to the generalised mobility that writing technologies accord, literature about travel has an additional mobility inscribed within its premise. It is a *guiding* literature, it takes the imagination travelling to specific places, casting the modern reader in a variety of traveller-guises, as Cunningham outlines above. It also sets up conditions for apodemic literature to do its performative work, instructing as to the best ways of enacting some of these roles. Depending on which guide is selected, according to which imaginative set of possibilities each tourist brings to a journey, we do indeed find guides that cast the tourist as Quixote or Crusoe, as detective, explorer, pilgrim, Romantic wanderer.

History of German Travel Guides

Baedeker guides, a German product, remain the oldest, most famous examples of the genre of the travel guide. MacCannell (1976) famously used the Baedeker Guide to Paris as a basis for his structuralist and semiotic analysis in *The Tourist*. He remarks upon the wealth of information provided by the guides, upon the role played by guidebooks in marking out tourist attractions as such, and demonstrating the way that 'modern society makes of itself its principal attraction in which the other attractions are bedded.' (MacCannell, 1976: 48). The guides also mark out their users and the relative 'quality' of their preparations, as we noted earlier, when using our own guides in public.

Another important aspect of Karl Baedeker's original guides was to release the traveller from reliance on hired servants and guides and; 'to assist him in standing on his *own* feet, to render him *independent*, and to place him in a position from which he may receive his own impressions with clear eyes and lively heart' (Cronin, 2000: 86). This point is clearly tied in to the competing ideological understandings of travel mentioned earlier. However, as Cronin notes, the omission of the question of language from this discussion is striking, given that 'the guide book translated the foreign culture into the mother tongue of the traveller. The traveller no longer had to rely on the oral translation of the guide/ interpreter as the guide book provided written translation.'

As we noted in Chapter 5, we found that the guidebook did not, ultimately, provide a significant release from either the need for translation/interpretation on the ground, or from the need to activate English or Gaelic, for tourists, on the Isle of Skye. The purchase and packing of guidebooks do, nonetheless, provide a release from the need to hire interpreters continuously, such transactions now being more fragmentary, for the purposes of guided tours around the whisky distilleries or the main castles.

The changing form and function of 'guides' – from employed servants to glossy mass-produced books – suggests that apodemic literature is a relatively modern form. The instructional use of travel writing goes back much further, in German history, than the production of the Baedeker guides, however. Bepler (1994) discusses the role of the traveller as author in 17th-century German travel accounts. Not only does her research demonstrate that reading travel literature was popular in 17th-century Germany, but she sees this genre as emerging in German-speaking lands and being based on the logical method of the philosopher Petrus Ramus. The application of this method led to the publication of travel handbooks containing detailed guidelines for the art of observation, particularly on note-taking and the production of classifications when travelling. The ideal travelogue, according to Bepler, was expected to provide instruction and moral improvement, but through observation, not through self-reliance:

> This projection of the traveller as an exemplary figure, rather than as a critically and emotionally responsive individual, determined the process of selection by which personal and subjective experience and anecdote entered the text. Travel itself was seen as a test of moral rectitude and a lesson in the vanity of human affairs, ultimately governed by the hand of providence. (Bepler, 1994: 184)

The role accorded fate in travelogues may thus allow us to regard these texts as *devotional literature*, a protection or a way of controlling fate, although, as Bepler records: 'By the end of the seventeenth century the providentialism [...], with its origins in pilgrimage reports of early sixteenth century, had degenerated into an empty formula and become the subject of satire' (Bepler, 1994: 193). 'Devotion' is a rather difficult term in this context, but we understand it anthropologically here, as defined by the OED; 'An offering made as an act of worship' – a kind of ritual insurance, investment against an 'act of god'.

Although the kind of devotion and structures of religion and ritual have changed somewhat since the 17th century, this is not to say that the role of 'fate', and thus of some form of devotion, is absent from apodemic literature. Anxiety and anticipation are not, as we have seen, absent from packing. Fate and self-reliance may be regarded as interwoven. Bausinger (1991) notes that rituals of washing, cleansing, eating and drinking are accorded greater space and time during the everyday activities of tourism. We have seen previously how washing and cleanliness are important aspects of the packing activities, and we shall return to these aspects in Part 3.

Devotion to the body, to its pleasure and protection, to the 'moral rectitude' of its actions, we may begin to argue here, become central concerns of tourism. Securing this pleasure and protection requires preparation and knowledge. When the hired servants no longer supply the water for the rituals of washing or the oral interpretations that keep us safe, the written translations of culture must do this instead. In this sense, we might augment an apodemic understanding of travel writing, with a notion of travel guiding as a form of devotion, an insurance against the unexpected and the unknown.

Travel Guides as Preparation and Devotion

German travel guides are begged, borrowed and purchased as part of the packing process. Leaving home without a guide is foolhardy, rather like leaving home with a bag that leaks. Airports are full of them, as are all the first-stop tourist information centres, the apodemics becoming increasingly specific and localised the closer the destination. Reading travel guides is part of the process of preparation and anticipation. It enables the imagining of the destination. It is future-oriented travel in the present.

Let us take the following empirical instance. In the main library, in a small German town, we were intrigued to investigate the shelves of travel guides for instances of Scotland. We found plenty, but perhaps most interesting to us was the fact that when we turned to the online catalogue, the following information in Box 5 had been left on screen by the previous searcher.

This is an instance of preparation and imagination. It is hard to say with any degree of precision what the preparation is for. We know not which of these books was desired, but this random search and classification does show the interweaving of different genres of text and different imaginary worlds. It is evidence of agency, of a limited but practical choice to direct some aspect of both financing and cultural baggage. It also shows, materially, the public funding of apodemic

Box 5 Screen in Stadtbücherei, Biberach an der Riß, Germany

> Braun, Andreas: *Schottland Handbuch*
> Chlupacek, Birgit: *Wandern in Schottland*
> *England und Schottland [England and Scotland]*
> *Die Highlands [The Highlands]*
> *Die Welt der Rosamunde Pilcher [The World of Rosamunde Pilcher – writer of fictions set in Scotland]*
> *Schiller: Maria Stuart [Mary Stuart]*
> *Maria, Königin der Schotten. [Mary, Queen of Scots]*

literature, both in general and specifically on Scotland, in Germany. These aspects, which may be drawn from this empirical instance, raise interesting questions. . Dominant lines of received questions would perhaps read as follows. Are we incapable of moving without instruction? Is our pleasure always to be prepared and defined by others? Must our travel, and, in this example, German travel to Scotland, always be prepackaged, categorised, domesticated, structured?

It is certain that such questions offer possible routes to potential answers, but for an understanding of intercultural aspects of material life, of the packing of cultural baggage, they are not especially illuminating. A different set of questions would ask: why might there be such a material, emotional need for cues, scripts, instructions and suggestions? What are the links between the anticipation of new worlds and experiences and the instruction manual? Why is there a need here for rules, pictures, tables, categories? What is about to happen to the body, and its spiritual, ritual and ceremonial dimensions that requires this kind of work?

Nelson Graburn (1978: 22) gives us one avenue of exploration with some initial answers to these questions:

> In spite of the supposedly happy nature of the occasion, personal observation and medical reports show that people are more accident prone when going away; are excited and nervous, even to the point of feeling sick; [...] Given media accounts of plane, train and automobile accidents, literally as tourists we are not sure that we will return. Few have failed to think at least momentarily of plane crashes and car accidents or, for older people, dying on vacation. Because we are departing ordinary life and may never return, we take out additional insurance, put our affairs in order, often make a new will, and leave behind 'final' instructions concerning the watering, the pets, and the finances. We say goodbye as we depart and some even cry a little, as at a funeral, for we are dying symbolically.

Through this quote, Graburn invites us to consider the fetishisation of preparation against the context of life/death relations. Our departures represent 'symbolic deaths'. Travel guides are there to anticipate and prepare us for the grief that is associated with the small, yet symbolically significant manner in which we must grieve for the things we shall leave behind, and steel ourselves for a future of unknown dangers. If our departures are symbolic deaths, then the cultural packing that accompanies the preparation for travel becomes the cultural and symbolic equivalent of, say, amassing grave goods. Weapons, broaches, shields, drinking goblets, charms and trinkets become insect repellent, money belts, travel guides, drinks bottles, much-loved teddy bears and suncream.

What we pack and how we pack it is symbolic of our preparation for this little death. What we read alongside the packing at home, and during the journey, is an aid to our emotional, material and physical preparation. It tells us something of the import of life back home, of nontourism, before we even leave. How much guides bought at home are actually used on the ground is a question to which we shall return in Chapter 8. Let us now turn to the guides themselves (see Box 6).

In order to gain a more structured understanding of travel guides as devotional, even morbid texts in the preparation for symbolic death, it is possible to consider them from several different perspectives; as texts that presume a domesticated or an exoticised afterlife; as texts that provide a quick transition through the change in daily routines, to the paradise of cultural plenty and of plentiful cultural and narrative capital; as highly, materially performative texts with moral instruction as to correct social relations; as sites for dreams, daydreams, Romantic understandings. Just examining the book covers of our guides already takes us in some interesting directions.

A place to sleep, places to eat, how to find places to eat and sleep, and places of pleasure all feature as the most important selling points of the guides. Consumers, interestingly, are not teased with promises of gift shops, souvenirs, or indeed of any portable, material consumables, these aspects come with much later apodemics, *in situ*. Instead they want to

Box 6 Wording and language on covers of different German guidebooks used

- *Baedeker: Schottland: Mit großer Reisekarte, 266 farbige Bilder und Karten. Viele aktuelle Tips, Hotels, Restaurants.*
- *Schottland, Architektur und Landschaft, Geschichte und Literatur, Dumont Kunst-Reiseführer*
- *Marco Polo, Schottland, Reisen mit Insider-Tips, Neu Jetzt mit Reiseatlas Schottland*
- *Freude am Reisen, Nelles Guide, Schottland.*
- *ADAC Reiseführer, Schottland, Ein ADAC Buch, Hotels, Restaurants, Pubs, Gärten, Festivals, Schlösser, Ruinen, Seen, Ausblicke, Kultstätten, Top Tips.*
- *Merian Classic, Kultur mit Genuß, Schottland, Land & Leute, Geschichte & Gegenwart, Künstler & Persönlichkeiten, KulTouren, Service & Sprachführer, M. Merian-Tips, Neu Merian-Karte als Extra zum Herausnehmen.*

eat, sleep, see and to be told tips, extra tips and new tips, as we also noted in Chapter 4. Those guides, such as *Dumont* and *Merian*, which suggest more, at the outset, than bed and breakfast, are presenting such intangibles as 'Land & Leute' (land and people), 'Service & Sprachführer' (service and phrase guide), not hairy haggis or Skye Batiks. Survival through transition, care for the body in the form of sleep and rest, and the tools of communication – these are the selling points for the guides.

Also of immediate note is the dimension of languages and intercultural communication. Out of the 81 words on the cover of our guides (see Box 6), 18 are English words, or words which are the same in both languages. Words such as 'Service', 'Tips', 'Classic' and 'Insider' are all Anglicisms which are more usually rendered in German as 'Dienst', 'Hinweise', 'Klassiker' and 'Eingeweihte'. We could use this as evidence of the erosion of the German language by the processes of English hegemony under globalisation, the privileged language of global capitalism. Perhaps we could read guides and their languages from within an understanding of globalisation as homogeneity, part of the rationalising process of a McDonaldised global society. Such an analysis speaks of the role of travel guides in institutionalising travel, a point noted earlier and, in 'routinising the charisma' of the yet-to-be-known, a reference to the Weberian roots of this discourse of rationalisation.

Importantly, though, the spread of commodities such as guides does not mean that there is a concomitant spread in the meaning attached to these commodities. In common with Wilk (1995), globalisation might be regarded as the spread of a common set of structures that mediate between cultures and through which notions of cultural differences are refracted. Perhaps it is at least as compelling to understand such language use, on the cover of a book that in itself marks a transition to the anticipated dream world, as a marker of that transition, an *interlanguage* (Davies *et al*., 1984), mixing lexis and syntax from two different languages. Equally, such language use could be interpreted as marking the much documented desire for authenticity on the part of tourists.

The discourses of tourism have only relatively recently received critical attention in the work of Graham Dann (1996) and the Cardiff *Leverhulme* project on tourism and global communication (Jaworski *et al*., 2003). Sociolinguistic perspectives have much to offer tourist analysis, but analysis of the role played by other languages, the learning of foreign languages, the choice of holiday destinations based on the knowledge that survival will be linguistically possible, these aspects have received little attention in the literature. This is perhaps surprising given that degree courses in tourism often insist on a foreign language component. The problem here is the confusion of functional language learning for the purposes of plying the trade and making a living, and the affective element of actually living in other linguistic worlds. Our travel guides all already demonstrate that survival will be possible, that interlingual worlds are already entered as part of the preparation and that translation is possible. They are, we may argue, reassuring as such.

There is an irony here, however, that is well stated by Cronin (2000: 86):

> With many others speaking your language the Planet is not such a Lonely place. Not only do the guide books deliver the travellers to the same places the world over but the language of the guide books creates a sensation of linguistic homogeneity. The independent traveller armed with the divining rod of the guide books finds by charmed coincidence that many of the other travellers in the guest house/hostel/café speak his/her language. This is hardly surprising if the same guide in the same language has brought the travellers there in the first instance. The autonomy of the printed guide (no local interpreters) produces another form of heteronymy (global interpreters dictating itineraries).

Interestingly, for the purposes of our argument here, Cronin uses the language of divination, of charm and of fate to describe, ironically, the work of the travel guide. He describes, again, guides that actually perform a devotional purpose, calming the elements, quietening the gods with their words and ways and offerings. The fact that it is actually no coincidence at all that we meet those who speak our language is, in its own way, an answer to our prayers. Linguistically then, as well as materially, the travel guide is a kind of ontological security blanket for survival in the unknown.

Packing Culture into the Travel Guide

We might extend our analysis of the social consequences of linguistic aspects of guides beyond the devotional and towards the *disciplinary*. In this regard, the common use of classificatory schemes and tables for presenting knowledge about destinations in travel guides is instructive. The guides do their work of classification, dividing up 'paradise' into manageable chunks, much like a children's school textbook with pictures, reminders and key facts; a veritable A, B, C (see Box 7 for an inventory of contents). Just as archaeological grave goods are suggestive

Box 7 Taxonomy of guidebook categorisation

Categorising Culture: A,B,C.... History Cultural Features Nature Biography Literature Scotland in cities, towns and regions – geography Town plans and maps – cartography Tours A – Z of information Phrase guide

of fighting, eating, risk and devotion, so the neat parcelling up of the destination is also suggestive of certain aspects of cultural understandings of paradise.

To travel without history, understanding of other ways of life, of festivals, languages, of geography, cult figures and literature is to be impoverished and potentially at risk of losing the investments of time, money and pleasure. Symbolic death to the modern tourist then, should not be undertaken without the resources of educational disciplines. The lesson to be taken from the apodemic literature is that lessons should be learned. What is interesting is that to gain the cultural capital of travel and experience, a degree of regression is required. The A – Z aspects of the guides are perhaps the most striking in this respect, evoking, literally, the lessons in literacy of childhood. To enter paradise, tourists should become like little children. The tourist body, in a different cultural and linguistic space, needs to begin its literacy learning again, for the kind of travel, that today is 'grounded in literacy' (Cronin, 2000).

A relationship between devotion, classification, knowledge and discipline therefore becomes possible. Such a concern with discipline and knowledge, understood through the lens of Michel Foucault (1991), extends the broadly neo-Weberian concerns with the institutionalising effects of Western guidebooks mentioned earlier. It does so through the insight that the disciplinary aspect of travel guides work at the level of the tourist body. To return to our earlier use of Bausinger, the relationship between instruction, devotion and discipline is one that takes form and consequence through the body of the tourist. It is a constitutive part of devotion to the body that characterises tourism. A concern with the political economy of the body was, of course, key in Foucault's *Discipline and Punish* in which he states:

The body is also directly involved in a political field; power relations have an immediate hold upon it; they invest it, mark it, train it, torture it, force it to carry out tasks, to perform ceremonies, to emit signs. This political investment of the body is bound up, in accordance with complex reciprocal relations, with its economic use; [...] the body becomes a useful force only if it is both a productive body and a subjected body. (Foucault, 1991: 26).

A way of producing disciplining knowledge that will make the tourist body a productive body, before, during and after 'paradise-Scotland', is by applying the categories of educational discipline. The tourist body has long been understood as an 'unproductive' body, as a body at leisure and the Catholic opposite of its Protestant binary: work. Our work suggests, however, that there is much to be done as tourists, even at the preparatory, anticipatory stage. We are being encouraged to enact the potential value of tourism by using the travel guide to discipline our bodies into action and thereby to subjugate it to calls of a wider disciplinary society.

Highlighting the devotional/fate-bound aspects of apodemic litera- ture are part of what Foucault terms 'the technology of the "soul" – that of the educationalists, psychologists, and psychiatrists' (Foucault, 1991: 30). These technologies, displayed in guides, are at work disciplining the tourist body with categories of knowledge it has learned to trust, so that it becomes productive and able to bring back narratives that have cultural power and become websites, albums, slide shows, travel guides and stories.

Conclusion

This chapter has focused on the instance of German travel guides and has examined some initial ways in which the German travel guide industry can be understood as more than simply another example of global capitalism at work. Koshar (2000) is very clear, in his own historical study of German travel guides, that the guides and their uses point as much to agency as to determinism. Travel guides are just one of the many different cultural items that are packed in the travel bag, and which are then unpacked and repacked during the tourist experience.

Packing, for the tourist, is not just about making belongings tightly secure, or filling a bag with belongings before departure. Packed objects have a symbolic, agentic status. Many meanings accrue to those items finally selected for the road. Obviously material items can be cultural but, metaphorically, we also 'pack culture' when we set off on holiday, and one of the prime sites for such cultural packing is the travel guide – where we still get our anonymous 'servants' to do this work for us.

The didactic and materially performative nature of travel guides has been our focus here, along with their part in preparing the body for a small, symbolic death. Packing is ontological: it is about our being as much as our knowledge. Objects such as tourist guides also have ontological status. Their *written* aspect lends an additional mobility to guides and to travel literature, which connects the modern tourist to specific places, but, importantly, makes these places detachable and re-renders them in the textual space of writing and literature. As such, in contrast to the oral narratives we examined in the previous chapter, written narratives – travel guides and travel literatures – act as less communal, conversational prompts to emotion, memory and intercul-tural communication. Instead they work with our imaginations, with a future orientation, that is more individual, private and personal, casting us in roles and lending us spaces for their rehearsal before we arrive on the tourist stage.

In addition, like oral narrative, the written artefacts of culture and those who own them may make claims on us as tourists. Travel writing will take our imaginations and our bodies to certain literary sites to *see* with our own eyes what we have *seen* in written stories. Oral narratives also took us to certain places, often more mundane and localised as the oral narratives we exchange as tourists led us to places, if not always to classified tourist sites. Travel writing – books – are also artefacts in ways that are distinct from oral narratives. We cannot pack speech, conversa-tion, performance. We can pack books, and we do. Packed objects, as we have suggested here, such as travel guides, possess a performativity, above and beyond their content. As material objects, they assert themselves on us and in so doing, may be ascribed, in analytic terms at least, a different ontological status.

We are aware here, as earlier, that it would be easy to construct a dualism between oral and written culture and that these two chapters may be seen to do this on the surface. A full discussion of the speech–writing, oral–written continua of communication and the debates which rage around them are far beyond the reach of this particular project. We are indeed making distinctions, but we are also pointing to gradations and similarities.

The buzz of experience and oral narrative or performance does not lead in any direct way to travel writing any more than travel writing directly impacts oral narratives, but we find the oral and the written forms in an intimate exchange, forging imaginative trails that criss-cross and leave their imprint the one on the other. We consequently find tourism as oral, written and as meta narrative, an energising, emotional, material activity that is generative of myriad, interwoven and inter-textual forms, and through which knowledge flows metaphorically as well as factually (Mcfague, 1975).

We are aware that, in highlighting these imaginative and performative aspects of travel guides, especially with our deliberate metaphorical use of a language of 'paradise', 'danger', 'symbolic death' and superstition, we are engaging in the tradition of understanding tourism as a liminal activity and that we follow on from work done by Turner and Turner (1978) and Urry (1990) *inter alia* in exploring the way that tourism relates to the structuring and anti-structuring of life and its ritual moments. In this respect, our preparations for departure instantiate liminal spaces of tension between the structures and anti-structures of 'home and away'. We shall return to this aspect again in the next section.

It is indeed possible to interpret the use of travel guides on the ground in a number of different ways, ethnographically and hermeneutically. However, in demonstrating the everyday nature of liminal activities, such as the packing of travel bags and the use of travel guides, we hope not to mystify liminal processes or to stress their separation from the ongoing work of living. On the contrary, the guides highlight aspects of nontourist life, our hopes, memories, expectations, discursive position-ing, our habits as tourists. Instead we highlight their ordinariness and their reliance on collective, cumulative knowledge and their place in the network of human and material life. In Part 3, we investigate the ordinariness of the human and material life of tourism.

Chapter 7
New Habits

This chapter redescribes and discusses the experiences of our journey and arrival at our first accommodation on the island and evokes the feelings of anticipation, out-of-placeness and dislocated habit of our initial few hours in this temporary new home. This rupture, to return to the Benjaminian notion mentioned earlier, has particular significance for our human bodies. These bodies are defamiliarised, out-of-routine, unsure of where and how, for example, they will wash, sleep and eat. We argue that such experiences are indicative of, what we may term, the problem of the 'unfamiliar body' in this particular tourist context. That is to say, the initial transition from domicile to destination involved in 'getting to' and 'getting in' a new place, involves the affective and material problematisation of the habitual and taken-for-granted rituals of the 'body at home'. We might label the latter the 'familiar body' with all its rituals, habits and foibles.

Rather like baggage, the tourist body becomes a tourist body only by dint of leaving home. En route to and whilst initially in this new place, the familiar routines of the body become temporarily suspended when faced with the demands of an unfamiliar milieu. We extend this argument by highlighting object-relations as crucial sites for the negotiation of the problem of the unfamiliar body. In this regard, object-relations, as defined in Chapter 2, take two interesting forms. On the one hand, they manifest an uncomfortable sense of the unfamiliar, as we shall see. On the other, object-relations become a central mode for negotiating this sense of discomfort and in making the new environment more familiar. Material objects, then, are crucial in the creation and negotiation of the unfamiliar body.

The final part of the chapter expands on how we dealt with our unfamiliar bodies through object-relations and a number of attendant social practices. We suggest that a conceptual labelling of this 'constructive' response to the problem of the unfamiliar body lies in the notion of the tourist's 'reinvention of the habitual'. This is never a finite act, but a continuous set of negotiations carried out in the context of emergent social relations. It is part of a process of structuring and anti-structuring, to borrow Turner's phrase. Using data from our first night spent preparing and eating food in the kitchen youth hostel, we show that this integration into the social life of the hostel, where we reinvent our bodies as familiar in this new provisional home, provided a fertile source for intercultural communication with other tourists. Let us start by 'getting there'.

The act of getting to a place is both a literal and a metaphorical transition, which opens the body to risk. On one level there is risk in the physical act of travelling whether by car, boat, aeroplane or by foot. On another, there is a different kind of risk and confusion emanating from the points where one gets closer to one's new and unfamiliar surroundings, that is where one begins to see the complexity of difference (different as in unfamiliar) 'up close and personal'. This is no longer a nontourist mode of being such as that described in Part 2. It is now an anticipatory mode of *doing* being a tourist. In these moments of transition represented by the journey, the knowledge acquired and prepared from the travel guide is put into question – those neatly packed categories of culture, demographics, local delicacies, are not all that useful when looking for the correct B-road to Uig and trying to connect the sometimes confusing details of an unfolding physical place with the tidy textual representations provided by a map. That careful packing of objects for the journey, maps and guidebooks, knowledges and advice for travelling, and the kinds of certainties associated with the definitions they create, cannot completely contain the unfolding largesse of the new space and time, even in these initial moments. Difference seeps out in a whole host of ways.

> We set out from Glasgow by car, heading up past Loch Lomond towards Fort William. True to its reputation, the road past the loch proved exceptionally busy and we were wary of the potential for a bump in the car. At this time of year (mid-summer), the roads were filled with tourists from all parts of the world struggling to cope with the volume of traffic on the tight and winding route past the loch. This sense of heightened awareness of both the potential dangers of summer driving, but also not wanting to miss a crucial signpost, meant our eyes were peeled and ears pinned back for the sights and sounds that would help us find our way to Skye safely.
>
> Our excitement about setting off to carry out the research, however, also made for other heightened feelings of trepidation, intellectual curiosity and good fortune at having the funds for, what for us at least, was interesting research. Our chats en route were fast, furious and focused, a combination of intellectual 'what ifs', interspersed

with childlike bleatings from our observations of cars with foreign number plates, or the remarkable number of T-registration hire cars. These cars were a sign that out research interests were near, that we were on the right track.

The initial excitement gradually transmuted itself into a quieter eagerness for our holiday. As the unusually hot Scottish summer began to take its effect on our conversation and our underarms, we tired of the long journey through winding roads, a feeling eventually eclipsed by the sight of the Skye Bridge. Something of a frenzied uncertainty set in, as we wondered, perhaps overanxiously, exactly where the turn off for the youth hostel was, where we might park and what exactly lay ahead. Such heightened forms of sensitivity (to potential accidents, to signposts, to visible 'research subjects') seemed indicative of a time where our ways of being and knowing, and their materialisation in our bodies, were at 'risk' in a number of different ways. That is to say, in this new context, all kinds of new scenarios were possible in which our relatively stable senses of self were opened and up for grabs.

Without even literally unpacking our travel bags, metaphorically they were already being stretched in all sorts of new directions by the uncertainties that lay around every corner. We had already begun to unpack, realising that things are different in ways that the travel guide as text could not possibly prepare us for. The risk to the categories of our everyday routines was prescient. The guidebooks did not tell us about the rules of the hostel, how it worked, who would be there and how we might negotiate it. The body was at risk then in terms of our habits and their negotiation with the unknowns of the hostel.

These questions about the unknown left 'dirty' symbolic and physical traces. They polluted the purity of our prepacked knowledges and expectations. They added layers of dirt and grime, exhaust fumes maybe, sweats (salty, watery physical sweat from the exertion of moving travel bags; smelly, nervous, urea-containing sweat from the tension). Transition takes its toll on our bodies as well as our categories; it opens the physical as well as the cultured body to risk. This interplays with the emotions that we experienced: bodily nervousness, excitement, disgust, pleasure. Perhaps travelling is tiring then, not just because of the physical exertions that often characterise the getting there, but also because the familiar physical, emotional routes and routines of everyday life have to be temporarily jolted and refound in new, unfamiliar places. What was once light (our baggage, our mood) had now become slightly heavier. No amount of preparation or awareness can change this reality for our travelling bodies.

As we saw in Part 2, our concerns with anticipation and the emergent sense of the provisional nature of tourist activity were anchored around objects. This was also true for our arrival. We might say therefore that object-relations became a key site for our emergent actions and feelings in our new accommodation.

> After checking-in at the youth hostel, for example, the first thing we did was to attend to our packed objects in the car. We had a conversation about the items that needed to remain in the car because they were 'valuable' and which might safely be taken with us into the hostel and kept in our rooms. Attributions of value were interesting here. Those items that remained in the car were our laptops and other pieces of equipment as well as our research monographs and some other documentation, which altogether accrued some significant economic value. We also left our boots and heavy clothing in the car. They would not be needed in the immediate context of the hostel, for they spoke of future walks and the predictable threat of Scotland's changeable weather.
>
> We only took our rucksacks containing clothes and washbags into the hostel with us. It was these that were most valuable since they allowed us to attend to our immediate personal needs in the hostel, not just in terms of the rituals attendant to our bodies, but also, as we shall see, for creating a more familiar habitat for us. It was at this point that a more obviously 'material' transition began to occur between stages of packing and unpacking. Having put our rucksacks down in our separate dorms, we headed out to buy a pint of milk for a cup of tea.

A sense of the provisional nature of our activities as tourists was inscribed in some of the key material objects in our new home. On returning to the hostel, milk in hand, we immediately headed for the kitchen. The fridge became a fascinating topic of conversation for us, as it seemed to index precisely the kind of time and space we were in as tourists. Later, we made a tape-recording of our reflections on the fridge, which included the following brief excerpt:

Alison: That fridge is fascinating actually with the name and date system.

Gavin: The names and dates on the food are incredible.

Alison: That's how transitory it is, that we're there, and that's how we mark our identity.

All the food in the fridge had to be marked with the name of the hosteller. This took us to the heart of our enterprise. Inscribed on perishable material goods was the provisional extent of the home. The names inscribed were, in the main, though not exclusively, German. Provisionality was more than just a feeling associated with being in an unfamiliar place. In this particular instance, it was also a social practice conforming to a set of unwritten rules about sharing the fridge and respecting other people's property. In this sense then, the items in the fridge exerted a kind of agency and acted as territory. They clarified to us that there was a certain way of doing things in this kitchen, and that we should learn this and follow suit. The object of the fridge dragged us into the social codes of the hostel, providing us both with material evidence of the provisionality of our home, and also the social means for negotiating it.

An uneasy sense of the unfamiliar was also manifest in other objects than just the fridge. Leaving the fridge, we decided to have an hour or so in our separate dorms. We wanted to have a shower (it was

Preparing food in a youth hostel kitchen

blistering hot outside) and some time on our own before moving into the kitchen/dining and reclining areas for an evening's research/ entertainment.

I, Gavin, spent quite an uncomfortable hour lying at the edge of my new bed before going to the kitchen to make dinner. I had an uncomfortable feeling associated with not being at home. Somehow this uneasiness, that something 'odd' or 'different', was in the objects around me. It was in their very smell, their very appearance – the mustiness of the beds, the lumpiness of the singular pillow, the cold of the shower, its anonymous pubic hairs and dirty footprints, the muffled noises from the other parts of the hostel. I wondered (well, maybe I knew) whether I was being slightly precious about things and whether I needed to 'get over it'.

Lying on the bed, I thought about how hungry and tired I was. Many questions passed through my mind. How were we to share this place with strange (perhaps in both senses of the word) people? How were we to share intimacy with others? Seeing someone else's stuff – who were they? Of course this was intriguing. Were they young or old? What nationality? How would we work around each others' bodies? Would they snore? Would they fart during their sleep? I consoled myself that only time would tell. Time would also only tell of the chances of getting food and quieting my rumbling stomach.

Meanwhile I, Alison, also disappear to my dorm. It is full of people, a veritable buzz of nationalities chatting away happily. I greet every-one, equally cheerfully, bag a bottom bunk and begin putting my sheet sleeping bag in place, correctly. I was buoyant. This was a familiar activity for me; I'd done this many times before in the past. My bags reflect this too. Out came the things I needed, all arranged to hand.

Above me is a Spanish child. She'd bought a Loch Ness monster, wearing a tam o'shanter and tartan, and it is sitting proudly on her bunk. Her friend is in the dorm next door and bursts in. I love the sound of the language punctuating the quiet. It immediately takes me places. They disappear up the corridor laughing. I start chatting with two women who are a little younger than me. They ask if I've been to the castle yet and offer me a leaflet. 'Is it expensive?' I compliment them on their English. 'I make lots of mistakes, but people seem to understand me.'

I go over to the bathroom for a quick wash. There are two women in there. One is having a shower and the other is talking to her and getting dried:

Soll ich die Schere mal einstecken von dir? [shall I put your scissors away for you?]

Mmm

Du, darf ich auf dein Schlapp Schlapps stehen? [Hey, can I stand on your flip-flops?]

Du darfst alles. [You can do whatever you like.]

Wenn du jetzt andrehst, hast du genau die richtige Wärme. [If you switch on now you will have it at exactly the right temperature.]

I was ready quite quickly, comfortable in my surroundings, enjoying the sound of a familiar tongue that I'd not realised before that I'd been missing, and feeling relieved to be back in a hostel. It felt safe, familiar, restful.

The questions of food and rest above seem indicative of a particular concern with the rituals of the body and of working out how we might be able to continue the habits we have for feeding and washing. Even for those few precious minutes, or even hours, our taken-for-granted routines of caring for our bodies had been disturbed. The habitual aspects of our lives had changed and we needed to re-create some kind of routine. Because we were out of routine and in a new place, it seemed as though time had slowed down and that a more conscious thinking and explicit questioning had been returned to our lives. This greater temporal appreciation of questions of the body was indicative of a time where our bodily relations had been placed in question. It was clear that other new arrivals were in the same state. Time was experienced as discontinuity, as rupture. It was a moment for change and a moment for the uneasy process of bodily reinvention. In this context, the problem was that of the 'unfamiliar body' – that is of the disturbances to the embodied kinds of knowledge that habit represents. In other words, the everyday habits of bodily attention that constitute the 'familiar body' would have to be altered. We would need to deal with an 'unfamiliar' body in unfamiliar object-relations. We would need to relearn habits in new places with old and new things.

As well as encoding the sense of the unfamiliar associated with a new environment, object-relations also played a different role. That is, object-relations were also crucial sites for negotiating and thereby reconfiguring the problem of the 'unfamiliar body', i.e. the body out-of-habit.

Returning to the dorm above, I, Gavin, took out of my rucksack objects that made this space feel a little more like it was mine, albeit temporarily – my alarm clock, the tartan boxer shorts and T-shirt which I wore at nights, a notepad and pen. These were all items carefully packed because of their centrality to my everyday life – to my routine. I placed these objects in and around my bed, mainly at the top near my pillow. Perhaps they were at their most comforting there, as they were in reach – at my fingertips and within easy sight and access, In any case, the mixing of familiar objects with unfamiliar ones became a method not only for rendering this space less uncomfortable, but also for marking it out as my own, however provisionally. This temporary sense of 'accommodation' again left me slightly as I took out my washbag and a towel, and went for a shower. Again – where should I put my towel? How did the showers work? Was there privacy? Would there be hot water – at what times of the day? Questions of attention to the body in this environment returned.

Even in these initial moments, we can see how object-relations took on two particular hues. Perhaps duplicitously, they were crucial both in the creation as well as the negotiation of the body-out-of-habit. Object-relations continued to be an important site for the ever-present negotiations that characterised tourist experience. Negotiation was never a finite act, but a set of continuous and ever-emergent social practices. We suggest that this constructive response to the unfamiliar body might be encapsulated in the idea that tourism involves a 'reinvention of the habitual'. It becomes instructive, then, to explore what such 'reinvention' looked like and the social practices associated with it. Such reinvention was, in important ways, carried out with other people as a social and intercultural endeavour.

Having got changed and decided that I, Gavin, had spent just enough time doing 'being on my own', it was time to enter the communal space of the kitchen in order to be around food and people. It was here that we actively tried to reinvent our routine in regard to food and people, and *we were gradually drawn into doing this within the context of intercultural relations*. Importantly, these relations took form in a particular and emergent set of object-relations and material exchanges. They needed a variety of phenomena – languages, people, stories, artefacts – to stake their claim on us and draw us into intercultural relations. Intercultural communication spun out of material needs and material relations to objects and took sustained form in dialogue. Tourism (understood as a reinvention of the habitual) was worked out in the demands of particular contexts (embedded), across and through the body (embodied) and through the practices of everyday life (everyday). Food, its preparation and consumption, was one such key context for tourism's embedded, embodied and everyday qualities. It appeared in both our journals.

> The kitchen is abuzz, but not noisily so. There is quiet concentration as we prepare our food. I, Alison, go over and pick up our pasta and instant sauce and begin to find pans, smiling at the lovely assortment of dinted teapots under the sink. I'm fascinated by other people's food. There is beautiful preparation and turns of conversation between people on the subject of its preparation as they work:
> Ach, so hab' ich das nicht gemeint. [No not like that.]
> Wir könnten ja ein Kochbuch schreiben. [We could write a hostel cookbook.]
> Someone is frying sausages, potatoes, onions and my stomach is rumbling. I want to stay in this space with these people, cooking like this, but we have been a bit too practical and only have instant-dinner, for ease. This was definitely a mistake.

Chopping a pepper in a youth hostel has complexities to it that might evade the habitual eye. This is because, when carried out in an unfamiliar kitchen with unfamiliar people, it entails a subtle set of negotiations around our bodies (as well as the bodies and languages of others), and their habits in the youth hostel. Of importance are the subtle aspects of moving the body; the nervous initial moves; the way that the repetition of actions firms things up in social facts; the trepidation.

> I, Gavin, went into the kitchen where someone was already cooking – a middle-aged man at the hob. He shot me a look. The kitchen had two standing tables at it surrounded by cookers, some work surfaces and a couple of sinks, and a fridge on the wall. I decided to use the table furthest from the man. Also in the kitchen were two girls who were involved in an intense conversation in a North American accent. They were colonising the other table, so I had to use the one near the guy. I took the peppers from the part of the open cupboard that had provisionally become ours (we marked it as such) where we stored our food and put them onto the table, as if claiming that it was my space and that no one else should use it. I did not know where the other chopping boards would be; I also needed to find a pot and pan. I went past the girls to look in a drawer to find a knife, but no chopping board. I apologised for knocking them as I went by. I then

saw a chopping board by the sink – I dried it and put it down. The knife did not work. It was blunt. I returned to the drawer to find a better knife. I by-passed the girls again, this time with a smile, an acknowledgement, that I wanted past. I then started chopping my peppers and I received another smile this time from the guy cooking at the hob. By this time, I began to know something about this space – its layout, which utensils worked and did not; I had worked my way into this space as a stranger through the repeated actions of glances and apologies. In a way things seemed slower, more intentional, more 'worried-about'.

New habits (in this case food preparation) meant new embodied, material practices. It meant fumbling, being clumsy, being vulnerable, being new. It involved a practical way of knowing how to 'do' the youth hostel: how to go about preparing food in this shared space. But it also involved a working out of how to be. It involved placing our bodies into the space, and demanding recognition for it, as much as it involved embodying the space and its symbolic, rule-governed order. This is not to say that the hostel was a space that somehow 'suggested itself'. Rather it is to suggest that its ontological qualities emerged through our interactions with others. We brought it into being through our relations with others. However, this should not serve to downplay the importance of the physically constraining aspects of the space in influencing the social relations that mediated our interactions. The physical layout and therefore the 'materiality', as much as the culture of the youth hostel, were responsible for making this particular context one in which communication between tourists was facilitated. The communality of the kitchen, unsurprisingly, enabled interaction around the process of cooking. We reflected on this and on the significance of physical space together later:

Gavin: There are five gas stoves I think, and there are four sinks, or five sinks as well. And there's a plate cupboard marked which is also an important division and of course signifier. And what was noticeable was that those all provide different locales, they're divided locales, in which it's more likely that you're going to have different kinds of interaction. So for example if you're standing at a cooker, it's more likely that you're going to share that space with somebody else than if you're standing (. . .).

Alison: By the little cubby hole.

Gavin: Precisely. And the preparation table's the same, these are important materials for humans which help frame the discursive landscape too. The other thing I was thinking was that ok, the material space, and the way that the room is divided up materially and socially is important for creating and facilitating different forms of interaction, but they also become cultured spaces in the sense of, the preparation table last night was occupied by four ladies who had North American accents and I originally thought to myself they were American but it transpired they were Canadian. So that became a kind of North American space where they were talking about their travels and there was one making fairly laborious comments on it. And then what we had was the stove that we were cooking on had the German guy cutting his potatoes, the father of the other guy was behind faffing about I think and doing things with salad and eggs.

Alison: And he was between the two of them, he was supervising what they were doing with the cooking as well.

The spatial layout of the kitchen, as much as its public symbolic order, had a material effect on us and other tourists. As a public space, its occupation and temporary ownership was always up for negotiation and this negotiation took the form of finding a place to store one's dry food, finding a place amongst the other things in the fridge and then working out where the cooking utensils were. So in terms of space then, its lack of clear ownership invited us into social relations in order to negotiate. It fostered the existential conditions, discursive and material, for particular forms of intercultural communication.

Across all the hostels we were in, food and the process of cooking proved a rich source of encounter. Those who shared our language and culture – in particular a family from northern England – commented to us on the way that other European nationals prepared their food. The monocultural activities became punctuated with 'other-cultural' language, reflection, food or objects. When people sat down to eat in the dining room there would be smiles of appreciation and approval shared. The encounter with other cultures in intercultural space, in a hostel on Skye, was welcoming, the space tightly shared with little possibility of cultural insulation. Having nipped out of the kitchen to write down some of the reflections about chopping peppers, we returned to cook our evening meal and to continue our process of research. There was a hum of different languages and different cooking processes – there was much potential for engagement. Some conversations were ongoing but we were not part of them – we were active observers and no more. How, we wondered, will engagement occur? When will the intercultural encounter move from the language of gesture and tacit acknowledgment into engagement and into speech and into other languages than English?

It happened for the first time, and then with regularity after the first encounter, through material objects and through material needs. We were cooking pasta and sauce. A German family was cooking on the stove next to us. We needed matches to cook and had to ask to use them – engagement began and a narrative exchange unfolded in German. We eventually ate together; they offered us salad and we offered them a willing ear for their stories. Perhaps inevitably, the conversation that ensued was one specifically about tourists in Scotland, what they had seen and done and whether their own material and cultural unpackings lived up to that which they had already anticipated.

Our encounter with this German family involved the reinvention of our habits for nourishing and sustaining our bodies. It grew out of a material need for food and materialised in the everyday practices of preparing and sharing food. It took form in the constitutive practices of exchange that saw us swapping ingredients and other basic foodstuffs for nourishment. In this respect, there was an interesting transformation of value involved. The foods we each had, had been procured from local supermarkets, part of a supply chain which determined the availability and 'value' of basic food stuffs through a process of capitalist equivalences and money-based transactions.

In the context above, however, there is a shift in the regimes of exchange through which the foodstuffs (lettuce, other pieces of salad and green tea) passed and a concomitant change in the kinds of values negotiated. The value being instantiated in this exchange was not an economic one, so much as a social one. It was one of friendship, a kind of hospitality and a sharing invitation to food. This exchange occurred 'outside' the market, in the emergent economy of the youth hostel. The lettuce was passing through different exchange regimes based on competing principles and procedures for exchange. Its value was not determined through the decontextualised equivalences of market exchange based on money. Rather it was a marker of a highly contextual exchange process enacting the kind of gift principles, which we reviewed in Chapter 2. What was being exchanged therefore was not just a lettuce, but also a set of glances, friendly smiles and shared understandings about blunt knives through which we came to 'break bread', both literally and metaphorically, with this family.

Moving from the kitchen to the dining area, our sustenance was one of stories, not just of foodstuffs. As we ate together, they told us about the walks they had been on, sometimes as recommended in the guidebooks they had. They gave us exact routes, looking vainly at us in the hope that we might know where these places were and be able to share something of their experience. They told us about the fantastic weather which they had not expected. They told us a little bit about where they came from. We were given a brief glance into their life. Perhaps the most animated part of this story-telling was around the Loch Ness monster. The father had some fun with us at his sons' expense as we all pretended to them that we had seen the monster. Some of this conversation involved a sense of comparison – comparing the expectations they had about Scotland, built up through myth, guidebook and everyday conversation, against its reality. Here, the father was freezing the subjective and emergent processes that constituted his experiences of touring Scotland into a set of narrative structures.

This was the first time we began to gain a glimpse into and understand the importance of narrative and story-telling as a key mode of exchange in intercultural tourist interaction. Moreover, it is important to recognise that this conversation ensued in and through the mixing of the English and the German languages. We all mixed and matched our languages as we pieced together a linguistic patchwork of conversation. In this conversation, intercultural communication flourished in and through languages, narrative and a variety of material object exchanges. We had begun the process of reinventing our habits. The acts of chopping peppers, borrowing matches, swapping glances and smiles, gifting lettuce, sharing conversation and drinking tea became important social practices indexing the practical ways in which we came to know more about and therefore feel more comfortable in this provisional home. As we reinvented our bodily habits around food and sleep, we negotiated and altered the embodied knowledges that enabled our relations to ourselves and to others. These were important moments of communitas.

Turner, in his late work on ritual and theatre (1982), identifies three forms of communitas: spontaneous, ideological and normative. In this book he is influenced by Buber's (1954) work on dialogism and on the Ich:Du (I and Thou) of interhuman relationships. Communitas is a form of social, collective being in which normal hierarchical structures are suspended, though importantly not erased, from consciousness. In short, he sees communitas itself as a characteristic form that emerges in times of anti-structure, when social rules are suspended. Spontaneous communitas is 'a direct, immediate and total confrontation of human identities' (Buber, 1954: 47). Ideological communitas refers to ways of conceiving of spontaneous communitas, the development of ideologies of such forms of social and human interaction. Normative communitas refers to the groups that attempt to foster or maintain spontaneous communitas on a more or less permanent basis, such as religious orders, for example. Communitas, for Turner, then is both part of the experience of everyday life, and a mode of being, a *habitus*, which has not forgotten the normal rules of the social game. This, we find in our examination here, is also the case with tourism.

Tourism can be a place of spontaneous communitas – the forming and forging of new sets of relations, unmediated togetherness, of alternative ways of doing the work of living, of spending, cooking, washing, talking, sharing. Tourism also has spatiotemporal dimensions onto which we project, in the past, present and future tenses of our imaginings and memories, an ideological communitas. And in the form of youth hostels, for instance, it represents a movement that promotes semi-permanent, normative spaces in which spontaneous communitas may be sustained, albeit with a constantly shifting population. Obviously youth hostels are particular spaces, lending themselves to the experience of communitas and of intercultural encounter (Richards & Wilson, 2004). We would not wish to claim the experience of communitas for all our tourist encounters or for all tourist experience, though we would argue that tourism offers a distinctive *opportunity* for the reroutinising of everyday life that can be radically affected by experiences of communitas, should they occur.

Communitas can be transformatory and narratives of experiences of communitas show that emotional, sensuous dimensions may claim us in the reroutinising of our communal lives as tourists, in new places. Communitas is indeed the things Turner defines it as, but under the liminoid and liminal conditions of tourism to Scotland it also involved connecting emotionally with the varied phenomena of touristic life. It involved the claims and the claiming of new people, new material objects, a range of new stimuli for the senses and our feelings, other exciting accents in these new places. And for this exchange of claims to occur the conditions of communitas are required. There needs to be a suspending of the normal rules of life and a sensuous mode of social being out of which new structures may emerge. There also needs to be a pause, a rupture, for the tourist to rest in a somewhat startling and transformatory reciprocity of perception, a sense of togetherness with abundant new phenomena that are being lived from within the attentive tourist body.

We went to our dorms that night tired, but excited. The bed that had earlier seemed cold and impersonal, now exhibited the personal signs of alarm clocks, dirty pants and a book. The tiredness meant we thought less about its ever decreasing strangeness. As I, Gavin, nodded off, I saw the co-residents of this dorm fumbling about in the dark, taking off jeans, talking quietly, sliding into bed. Some whispers to each other. A fart. Some muffled snoring ensued as my dorm mates, who seemed to have been here for longer, slept.

Chapter 8
Exchanging Stories

In the last chapter, we evoked the provisionality of tourist life as a material phenomenon centred around the problem of the 'unfamiliar body'. We argued that getting to and getting in a provisional home was a time during which the body, as a material, emotional and cultural artefact, was at risk. The habits of the body had become disconnected from their familiar locations, and new and practical ways needed to be found for living in this new home. We explored the duplicitous role of object-relations in creating, but also mitigating the feelings of discomfort associated with our 'out-of-habit' bodies. From this perspective, we suggested that tourism could be viewed as a social activity involving a 'reinvention of the habitual'.

In the latter part of the chapter, using the example of preparing and consuming food, we showed how this reinvention took form in social practices through negotiations with other hostellers. Our insertion into the social relations of the youth hostel involved an intimate engagement with its material spaces (the spatial division of the kitchen), material objects (matches, spatulas and peppers) and material practices (cooking, smiling, eating) through which spontaneous communitas was created and intercultural communication enabled. Talk, in its multiple forms, was generated by and generative of such different materialities and a key mode for spontaneous communitas. It is these different 'materialities' associated with tourism and intercultural communication, and their associated emotions and language, which provide the point of departure for the present chapter.

It is our aim here to develop our explorations of this 'material' base and its attendant emotional states in the context of the flourishing of intercultural aspects of tourism. In this regard, we unpack how inter-cultural communication flourished through a variety of exchanges with particular emphasis on the fact that such exchanges occurred in times of spontaneous communitas in language and through narrative (as high-lighted by the examples of our interactions with the German family in the last chapter). Again, we should stress here that our instance involves the particular context of youth hostel tourism in Scotland and that we are not attempting to provide a grand theory of tourism as communitas. We are concerned with certain intercultural, touristic phenomena and their reciprocal play in our perceptions, in a particular context.

Exchange and narrative – oral as well as written – we argue, were key propagators of the moving cultures in which intercultural, touristic phenomena took form. In this regard, we show how the relationship between social practices and stories constituted a key dynamic in intercultural exchanges. Practices led to stories, and stories led to practices. This is a kind of hermeneutic circle for intercultural tourist communication, in which practices and stories provide the conditions of possibility for each other.

Inspired by Lury (1996) and Turner (1982), we substantiate this mutual ontology of practices and narratives from two analytic perspectives. First we investigate how the 'social life' of tourism and intercultural communication *'has things'*. In other words, we continue our line of thought developed in the last chapter by exploring how tourist life is characterised by a variety of materialities (objects, practices, bodies) that often lead to intercultural encounter. We develop this line, in more specific terms, by focusing on how the diversity and creativity of such material and emotional exchanges led to particular kinds of stories. We explore the important social, temporal and formative roles of stories in everyday tourist life. Second, we investigate how certain 'things' and 'feelings' in the domain of tourism have a 'social' and thereby an 'intercultural life'. In this regard, we explore the 'biographies' of travel guides, ours and those of other tourists, as they circulated through various social contexts.

Everyday Tourism: All Play and No Work?

As we consider the way tourism as intercultural communication *possesses* both things and feelings, we are drawn to consider the work that occurs in the reroutinising of everyday life and in the doing of tourism. The work of tourism we see as enabling a certain possession – of things and of feelings – but not in the narrow sense of consumer culture alone. It is crucial, at this juncture, that 'work' not be understood in the restricted sense of paid work, but more widely, perhaps as task or as commitment to an endeavour. Just as it is crucial that exchange is not seen to occur only in capitalist relations. For us to speak of *the work of tourism*, of the having of things and feelings that come to us as a result of participatory actions in the world, a wider definition of work is required than that which dominates both sociological literature and popular understandings in the West.

In other words, and this is a key point, our understanding of the *work of tourism* requires us to reach out beyond the tropes of the alienated modern and the Romantic or exotic idealised other to a sense of work as encompassing rest. For us to argue, as we do, that tourism creates opportunities for the rewriting of the rules of everyday life, for a certain embodied reflexivity and a privileging of the imagination in the West, we are drawn to reconsider the binaries of paid work and paid rest.

This sense of tourism as work is perhaps best (although as we shall see in the next chapter not exclusively) seen in youth hostels. Hostels were an important site of creativity, and self-creation at the heart of living, of life, cooking, washing, cleaning. People sharing space and time, negotiating values, sometimes effortlessly, sometimes uncomfortably. It was here, in the midst of the labours that this involved, in this nonprivatised spontaneous communitas, that, as we argued earlier, intercultural communication flourished. The creation of any community, it would seem, requires a hard, common task (Ferguson, 2001). Its structures of exchange are material figurations around everyday needs, practices and objects. Tourism as intercultural practice, then, cannot have a value divorced from everyday life and its embodied materialities, and the feelings this engenders of renewal, labour, cultural change and dislocation, to name but a few.

As we suggested in Chapter 2, the work–leisure binary is a frequent trope deployed by various kinds of researchers to structure a particular understanding of tourism as one where the body – its functions and emotions – are at rest. Such a distinction is problematic. For one, it is too simplistic and clear-cut to suggest that tourism is all leisure and no work. In Box 8, we have listed the kinds of everyday activities in which we engaged on our holiday on Skye. These are indicative, contextually specific forms of exchange that go beyond commodity relations and instantiate a broader outworking of both work and of possession.

We would argue that this table blurs the line between tourism as work and tourism as leisure. We continued to wash ourselves, to feed ourselves, to meet new people, to dispose and dispense of things. These activities made us *feel* better. For us, these might all be counted as forms of work, in this case indexing a certain care for our bodies, and a certain labour in the form of social interaction with others. The problem, to repeat, is that 'work' has been too narrowly confined to the notion of 'wage labour', key to the capitalist labour process and curious to the industrial and economic development of certain Western societies.

Extending the notion of work beyond the confines of the wage-based economy, we can view the reinvention of the habitual domain in tourism

Box 8 Practices of everyday tourist life

We observed and participated in the following:
Washing
Cooking
Disposal and dispensing
Defecating
Sleeping
Tidying up
Telling stories
Shopping
Gifting behaviour
Spending money
Visiting attractions
Meeting new people
Preparing food and sharing food
Recording practices ˙
Exchanging
Speaking foreign languages
Engaging in intercultural communication

as a key site of the 'labour' involved in touristic, intercultural exchange. This labour works with and through objects and languages. Box 9 presents a brief inventory that gives a few instances of what we observed and participated in exchanging.

Even this most brief of inventories demonstrates the diversity of things which tourists might be involved in exchanging. Whether nectarines and postcards, tea or toy animals, the everyday material realities of tourist exchange implicated a whole host of different and sometimes surprising objects. What this means, then, is that it is too simplistic to understand tourism merely as a form of leisure based upon a series of money-based exchanges circumscribed by some kind of abstracted and anonymous 'market'. To repeat, tourism is not just a consumer industry – it is industriously consuming of self and other. Indeed, we might also attend to an understanding of tourism as an embodied form of 'work' (or labour) involving a multiplicity of different forms and objects of exchange. A notable form of exchange involved the use of different languages.

Talk about Talk

Where intercultural encounter and interhuman acknowledgement occurred was in such places of life at work, of 'essential functions'.

Box 9 Objects of exchange

Nectarines
Tee mit Milch
Shortbread
Croissants
O-saft
Stories
Personal histories
Drinks
E-mail addresses
Telephone numbers
My favourite things: Whisky, music and literature
Culture-specific problems: Ostdeutschland
Die Wende: Our memories
Politics and socialism: Quotes from Stefan Heym
Ostprodukte
Car lifts
Walks
Photos
Scenery
Viewpoints
Careers and aspirations
Studies
Languages
Midgies
Gaelic
Memories from France
Single malt
Guidebooks
Films: The Full Monty
Jeremy and Felix
Postcards
Stamp prices
Where to go
Mallaig ferry times
Gift ideas
Weather

This might be a kitchen; or it could even be, as one of us found out, the women's toilets. Here conversations were exchanged between different nationalities which engaged with the material circumstances and spatial arrangements. Comments such as 'it's just too narrow in here – I cannot get through with my rucksack' – and gestural engagement in confined spaces led to other comments on the weather or the place. Material objects, confined spaces, the plethora of signs of culture and of potential interculture were foci for engagement.

Whether engaging with landscape, material objects, food, space, with people or their languages and signs of their lives, moments on the road took shape in language and in oral narrative. The role of languages in engagement was crucial. Languages functioned in a variety of ways as sense was made and communicated of experience and as narrative was born of the engagement.

For instance one of our observations concerned the mixing of languages (English–German and English–French). An example of this mixing of languages around material objects came in the Kyle of Lochalsh in a tourist information office where two Germans arrived and commented to each other: 'Wir sollten ja bald ein Pub aufsuchen (we should look for a pub soon)'. Certain cultural objects made a mixing of languages possible because of the human–material interaction, the reciprocity of perception. Languages transform themselves around other-cultural objects. They were highly accommodating of new words and objects. We observed a similar patterning of languages in inter-cultural interaction.

After four hours in the same confined space of a boat trip into the heart of the Cuillins and Loch Coruisk, an international mix of day trippers returned to the landing stage in Elgol:

Teenage daughter:	Right, I'm going to get my French going again now.
Mother:	We're going to France for our holiday next year.
Daughter:	I'm going to write to her and get to know her ... what's birthday in French?
Mother:	Anniversaire?
Daughter:	Quelle date est ton anniversaire?
Mother:	I'll get those French books out.

A French and English family had clearly made an acquaintance and the most immediate need for narrative, following the trip and the parting on the landing stage, concerned languages as the most compelling object of conversation, not the incredible beauty and rarity of a trip into the Cuillins on a day of hot sunshine and clear blue skies. Ian Reid (1992: 1) maintains, as already noted, that:

We crave narrative and we crave exchange. Both compulsions seem inherent in human culture: to interpret our experience as story-shaped and to interpret it as reciprocally transactional.

Where we stayed for two nights in the hostels, we engaged with other travellers and were able to exchange stories. At the end of the day we would meet again, having experienced something of the island and this experience would find expression in the stories of the day – shared in the kitchen, over washing up, in the common room or the dormitories, or after the meal over shared cups of tea or coffee. Sometimes these stories were in English, sometimes in French and German. Our facility with other languages led to much discussion of those languages and language learning. Where Gaelic was spoken as the first language, we stumbled through our own tourist phrases, sharing these with others later, seeing them reappear on postcards and in journals as proud tokens of their attempted engagement. Some read their guidebooks, others spent the evening writing journals or postcards as further expressions of a craving for narrative and exchange.

One of our observations concerned the extent to which other languages were present as an aid to travellers. How much translation was used in signage, hostels, tourist information and hotels, for instance? In the main there was very little and most of those in the tourist trade had only English with some also having Gaelic. Where translation was used, it featured margins and as a protection. Signs saying 'The cliff is very dangerous' or 'drive on the left' were translated, as were instructions to women on how to dispose of their sanitary protection.

In the main the translations were poor, unless commissioned by the Tourist Board, where they were bland, because correct and promotional. Where translations had been attempted by individuals, they even tended towards the hilarious, as Box 10 demonstrates.

There is evidence in this box of a clear desire, on the part of the hosts, to reach out in hospitality and in response to what we have already identified as a key context for intercultural communication – that of food. Languages are offered to us in a willing spirit of communication. They serve to entertain through their mistakes and are received with laughter, becoming our favourite story, a souvenir or trophy from the road.

The fact of story-telling hints at a fundamental human unease, hints at human imperfection. Where there is perfection there is no story to tell. (Ben Okri, *Birds of Heaven*: Aphorism 18.)

Drive on the left. Road signposts in the highlands

We found a paradox at work here that is mirrored in later chapters and also reflects our earlier observations around oral and written culture. This paradox involved the richness of the imperfect, rough, unpackageable aspects of touristic life as contrasted with the perfected, professionalised, even expensive, luxury of glossy brochures or plush hotels. Poor translation claimed us. Correct translations were immediately forgotten. Here, at the interface of languages, we found a space opening out for communitas and for a sense of emotional connection with languages, not possible when perfected.

Box 10 How would you like your eggs?

Eggs-poached	Oeufs à braconné	Eier-hat gewildert
Scrambled	a monté	hat geklettert
Fried	à fait frire	gebratene
Boiled	Bouilli	hat gesotten

Thick with Stories

Asking questions about the role of languages and translation in our experiences of intercultural communication on Skye brought us, then, to the heart of narrative as a key form of tourist exchange and therefore as a key modality for 'doing tourism'. The act of story-telling had important social, formative and temporal functions for tourists, which we now explore.

The milieu for tourism on Skye is thick with stories. Stories of the past, of old ways of life, of the whisky industry, of clan battles, of Bonnie Prince Charlie, of the Highland Clearances, of a Gaelic Renaissance. Ceilidhs for tourists told such stories of the past through music, dance and narration. Visitor centres provided a visual and aural narrative for tourists and locals alike which explored the history of Skye through a number of lenses. Shops sold a piece of these stories to take home, through guidebooks, glossy historical pamphlets, clan key rings and overpriced videos.

Souvenirs represented the objects where stories and emotions could find focus and possession, not simply as a form of commodified relations but as objects that had a wider symbolic claim and wider narrative potential. There were plenty of official stories, part of the wider written culture and the mythic mobility and public knowledge flows of Scotland, from Scotland and about Scotland. These were stories commodified mainly for the tourist market, but also for locals. They were 'official' 'written' cultural stories (Turner, 1982) if you like, part of the wider public script that gives form to the island as a tourist destination and heritage site. These resemble the stories discussed in Chapter 6. Of equal interest, alongside such official stories, were the oral stories that tourists told each other in a variety of different contexts and were discussed in Chapter 5. Our journals indicate this.

We met a family from northern England – mother, father and two children – who had been to the island on several occasions before.

This time, they had been there for a week already. We chatted to them
enthusiastically on our first night in the Bluewater hostel and they
began to tell us about some of the things they liked to do with the
children on the island, why the island was a great place to take their
kids on holiday and what they would recommend that we did. They
told us about the weather and how they thought the island was great
for the kids to get outdoors, get some fresh air and pursue long walks
in the mountains and on the beaches. They stayed in youth hostels
not only because it was cheaper than a B&B or a hotel for a family of
four, but because it meant that their children would come into contact
with people 'from different cultures, who speak different languages'.
They would learn to share time and space within intercultural
relations.

This part of our conversation contained interesting formative aspects
regarding normative communitas and its transmission, through the
emotional and material impact of varieties of intercultural exchange.
They were negotiating value for money – not being in a Bed and
Breakfast, nor an expensive hotel allowed them to have a cost-effective
holiday. Their sustained and particular comments about their children on
holiday and the opportunity to meet people from other cultural and
linguistic backgrounds suggest, however, that there was more to their
choice of accommodation than mere thrift. Economic choices, contrary to
most psychological models of consumer behaviour, are not a matter of
isolated and abstracted rational criteria. Other important values come
into play. On one level, they were perhaps instilling an interest in their
children in other ways of life, as well as a certain respect for alterity to be
gained through the experience of being together with strangers.

Their emphasis on the experiential and therefore embodied and
affective knowledge to be gained from exposure to others in a tourist
context tells us something about how these particular parents encouraged their children to learn to be tourists, to see the landscape and to
interact with others. The children were being inculcated in particular
material practices as tourists, which introduced, supported and sustained particular ways of experiencing the world. There is almost a rite of
passage here into a world seen through the lens of the parents. Perhaps
one might also regard this kind of formation as indicative of the subject
positions of the parents as a white, middle-class family. The values
enacted as part of these intersecting positions seemed to guide their
holiday practices and their practice of narration – the what and how

something is told. This family instantiates the 'relatively repetitive social system' of "normative communitas' that Turner discusses (1982).

The formative element of their story-telling related to us as well as the children however. They recommended that we find and spend time at the 'coral beach', a sandy area beside a loch somewhere on the centre of the island made from crushed and sedimented coral. We knew this must be a special place. Their revelation occurred after a long period of chatting together, of building up a list of suitable places for people 'like us' to see. They told us in a hushed tone, as if they were imparting a secret to us that only they knew about. It was 'off the beaten track. You'll not find it in any of the guides'. They had been told about this beach by a set of German tourists. They were passing on this information to us, educating us in what would be a good thing to do.

Time and Space for Stories

As well as the social and the formative aspects of their stories of the coral beach, there is an interesting temporal element to this mode of tourism that might be dwelled upon. The focus on time is interesting because it allows us to gain an understanding of how, through language and narrative, the idea that we are or have 'become' a tourist is achieved. In this respect, we see how tourism becomes an effect of social action rather than some kind of obvious *a priori* category.

At the heart of the creation of ourselves as 'knowledgeable tourists' is a shift in the time of that which we experience and a concomitant shift in the (verbal) tense of the language that comes to be used to express it. The first time we experience something on holiday, be it a night in a youth hostel or the discovery of a coral beach, our attempts to make sense of what occurs open us to the uncertainties, ambiguities and unfoldings of those moments in the present. We cannot yet say with any great certainty what it is we will eventually 'know' about this new event, as we are still making sense of it. Through time, and the repetition of our actions in this place, and then in narrative, we become more familiar and our knowledge forms more clearly. Tourists, we may argue, tell stories as a way of showing that they have been on the 'right track'. Through these stories they display the connections between place and emotion, the claims these make and the imaginative trails that blazed in narrative exchange.

After the second time of seeing and experiencing, we can create anticipation based on retrospection (Cooper & Law, 1995) – the likelihood of what is to come based on what has already happened. As emergent feelings and interpretations are turned into knowledge through the repetition of action and sense-making in language, there is a concomitant shift in the tenses of our experience reflected in the tenses of our language. The present turns into the past; the unknown into the known. We might say that this firming up of knowledge through the repetition of actions creates a kind of 'grammar of experience', which allows an event to be talked about through language, and its constitutive speech acts.

In the case of the coral beach, the family's repeated visits to it and probable multiple recountings of its existence to others like themselves, mean that they now consider themselves to have knowledge of it. The coral beach has moved from an unknowable present to a more knowable past. It becomes part of their stockpile of narratives, which can be drawn upon to engage in social action. The narrative of the coral beach, along with a number of other narratives this family has garnered over time, is part of what we might label an *emergent tourist grammar* through which we create knowledge of ourselves and our actions *in situ* as tourists. We witness here the creation of a narrative structure-for-action.

We do not wish to suggest that it is only through textual, grammatical procedure that we do tourism. This would be to 'commit scholastic fallacy' (Bourdieu, 2000: 53) of, to repeat, 'imputing to the object the manner of the looking'. Trained as we are as scholars in reading and writing, we are likely to project reading and writing onto all that we encounter. Identifying this as a direction in our thinking does not, however, mean that we should elide this with a call to eradicate all such modes of seeing and of sensing from our labour. Awareness of difficulties does not necessarily require an academic knee-jerk, rather more an acceptance of the limits of our endeavours and a way of working with the sentient and sensing human body. Story-telling, like research, is an oral process conducted in and through the body, its flesh and its feelings, its traditions and histories.

In temporal terms then, we might understand two different kinds of 'stories'. First there is the kind of story-telling that is often associated with the first time that we tell of a new experience. It is rough, ready and imprecise, rather like chopping peppers in the youth hostel kitchen with a blunt knife. But it is also key in making sense of that which we have experienced. We might call this 'story-making'. Second, and by something of a contrast, there is 'story-telling' or 'narrating', that point where a tale has been firmed up, tidied up, 'written', devoid of elements unnecessary to its unfolding. The case of the coral beach was one of story-telling or oral narrating. The difference here is one of time, a shift from the present to the past. Perhaps as tourists we tell ourselves stories with an anticipated future audience in mind, one that will be impressed by our tales of beautiful weather, stunning scenery and value for money. In any case, time was key in the creation of tourist knowledge. Narratives provided a grammatical structure for immediate social use and the creation of social distinction through 'restored behaviour' (Schechner, 1982). Through this emergent grammar, events on holiday cease to be bound by any immediacies of time and space. They can be continually renewed and re-created through narrative.

Stockpiles of Stories

The manner in which tourists stockpile stories provides interesting analogies with the role of the ethnographer in the field. Just as ethnographers immerse themselves in a cultural milieu in order to understand its systems, rules and scripts, later to redescribe these through various academic narrative conventions, tourists, like the family from northern England, make similar moves. In the particular conversation above, they too were telling tales and passing on advice on the cultural milieu in which they had found themselves. They were passing on their readings of various locations on the island, revealing its hidden aspects to us, as newcomers whom they had seen poring intently over the pages of our guidebooks. They positioned themselves not only as 'concerned and interested experts' however. They also cast themselves in a subversive role by suggesting the superiority of particular locations and experiences that were not in the guidebooks or on the 'official' tourist trail.

Not having to use the guide, it seemed to be suggested, represented the pinnacle of tourist-knowledge-in-practice. Here was a subversive narrative economy, which gave prescience and possibility to what the literature describes as the 'post-tourist' (Ritzer & Liska, 1997). The written guides were redundant, implicitly positioned as dulling the senses and imagination of the traveller. Oral narrative was the queen. The parents were touching and feeling the island semiotically, challenging what they believed to be the dominant script of the Isle of Skye, circumscribed by the interests of capital and offering an alternative feel. This narrative economy was a key part of what we might call a tourist ethnographic imagination.

This ethnographic imagination was certainly not confined to tourists. In the guest houses and B&Bs in which we were accommodated, the stories of the owners were also a key mode for social interaction. In one guesthouse we had informed the owner before our arrival of our intent to carry out research under their roof. We described to the owner exactly what we wished to do. On arrival, the owner immediately remembered who we were and why we were there when we announced our names. She was fascinated by what we wished to do and was clearly keen to facilitate and participate in events in some way.

An outpouring of different stories ensued from the owner. Stories of all the different nationalities she had accommodated and cooked for, the languages they spoke and the particular demands for food that they had. She told of the humorous side of international tourism, as well as some of her frustrations with certain nationalities. Her ethnographic imagination was clear to us, however, when she began to tell us of the plans she had hatched to enable our data collection whilst staying in the guesthouse. She was a fund of practical knowledge. In addition to the stories she was giving us as data, she was going to make sure that we took breakfast at a particular table in the corner of the breakfast room where she could place tourists of different nationalities around us. And this is exactly what transpired. We were seated the next morning beside three German ladies getting excited about the prospect of having some haggis for their breakfast. Our short stay there was characterised by the owner remembering experiences of intercultural communication that we might be able to use in our research.

Intercultural Friction

A further instance of a hierarchy of oral over written culture, a hierarchy upending the Western everyday norm, came to us when we were staying in a hostel where there was a sizeable number of young Germans who had different kinds of travel guides. Mostly in the evenings in this youth hostel, we would sit around the lounge area – a large public space in this hostel – and join others writing postcards or journals, playing with rolls of film or digital cameras and, importantly, either rifling through guidebooks or having them within finger tips. Here it would seem that the guides were playing their role as security blankets, as instructors, as preparation for the next day and as a devotional tool against the yet-to-come; it provided order against the lessening disorder of the island, as we discussed in Part 2. We met two eastern German women, who seemed to be less than convinced by now about the security blanket function of the guidebook they had:

Wir haben eine Holländerin getroffen und sie konnte es nicht vorstellen, daß wir ohne *Lonely Planet* reisen. Wir haben kein

Baedeker. *Lonely Planet* gibt's nicht in Deutschland. Ich hab *RoRoRo* dabei. Es ist furchtbar.
[We met a Dutch woman and she just could not comprehend that we were travelling without *Lonely Planet*. We haven't got *Baedeker*. You can't get *Lonely Planet* in Germany. I have a *RoRoRo* with me – do you know – it's terrible!]

Compared to the implied superiority of the Baedeker, the two women, let us call them Anna and Katrin, only had the RoRoRo guide, another slim youth-market pocket guide, one which was clearly failing in its instructional role. This conversation about guidebooks continued and led us into some revealing identity-work on the part of our German companions. For in this same evening, we saw how guidebooks could be used as a vehicle for social differentiation, and could explicate some sociological specificities of German travel to Scotland.

We were sitting with Anna and Katrin when a group of Germans all emerged from the dorms to prepare themselves for the next day on the road. They were poised with their guidebooks literally in hand, with keys for their hire cars, mobile phone in the pocket of their jeans and chatting loudly. The public space of the lounge had for a short time become theirs as they prepared themselves for the next day. At a guess, they were probably students, but well dressed ones and certainly not struggling for disposable holiday money. They were from the western part of Germany, probably the south – their dialects told us this, and Anna and Katrin confirmed it.

Anna and Katrin's reaction to this group was clearly uncomfortable. They became quiet; they glanced at each other a lot; they looked disapprovingly, eyes rolled. Our conversation became discontinuous, interrupted by their awareness of the other German tourists. Both Anna and Katrin, in emergent agreement, commented that these were 'Wessis' (slang German term for those from the former Federal Republic) which they identified from their accent and way of speaking as well as their clothes. It became clear that they resented this other group of Germans, although this was never completely explicit, but subtly encoded in their silent stares and at times, declarative statements about 'Wessis'. It was in relation to guidebooks, however, that they felt able to vent their spleen.

These 'Westerners', they felt, were overly reliant on their guidebooks, following them like sheep and not really seeing Scotland for themselves. Anna and Katrin seemed to be suggesting that they were blindly following these guides, and that this was 'typical of the Wessis' whom they appeared to be reading as victims of capital. Perhaps implicitly then they were following a line of thought, common in the resistance of the West, by Eastern Germans, where these West Germans were positioned as the real 'passive dupes' of consumer capital, and guidebooks just one instrument for this. In this trope, they were, we might argue, repeating the same kind of hard dualism concerning economic capital in the tourism literature, outlined in Chapter 1. Of course this implicitly meant that the judicious use of guides, that is use with a critical awareness of its potentially institutionalising nature, provided a morally superior way of travelling for them. Reflexively enacted in this way this is an important source of cultural capital for the post-tourist.

This instance of 'othering', of positioning a group in negative discursive relation to the self through the selective use of images and representations, is very much situated within contemporary socioeconomic concerns in Germany. It speaks of a context in which, contrary to German government hopes, the economic divide between the former east and the former west has risen sharply since the free market was introduced into the east, and the notion of a cultural reintegration is proving problematic. In a sense, what we might have here is German class relations of a sort, and a working out of domestic identity issues in a foreign context. In this respect, the guides very much take on the specific sociological hues of contemporary German post-Wende economic relations. This demonstrates how we bring emotional and cultural baggage with us on holiday and how the kinds of identity-work we do at home, we also do abroad. We do not leave our emotional baggage at the check-in desk, the ferry terminal or the railway platform.

As this particular example demonstrates, whilst youth hostels are sometimes terrain for the forging of intercultural friendships in travel, there is also potential for social friction and conflict, exclusion and division. Because of the ever-changing number and type of traveller, youth hostels become a site open to a wide range of competing symbolic orders and different ways of understanding and inhabiting this space ideologically. These are precisely the kinds of friction that Turner (1982: 50) describes when discussing normative communitas:

> Communitas tends to generate metaphors and symbols which later fractionate into sets and arrays of cultural values; it is in the realm of physical life-support (economics) and social control (law, politics) that symbols acquire their 'social-structural' character. But, of course, the cultural and social-structural realms interpenetrate and overlap as concrete individuals pursue their interests, seek to attain their ideals, love, hate, subdue and obey one another, in the flux of history.

Across the different youth hostels we stayed in, we met white, middle-class families; large groups of travelling Germans; a large Italian family; and backpackers of all sorts (French, Iranian, Danes, American, Australian). All brought multiple frames of reference, and therefore conflict potential, with them, which they evoked in each other and in us, claiming us all in different ways. With regard to the example above, this is a clear instance of 'intracultural' communication that dispels any myth that distinctive national cultural groupings will travel and inhabit tourist spaces in culturally homogeneous ways.

We might, however, also venture an interpretation of this instance in terms of the devotional aspects of apodemic literature mentioned in Chapter 6. As children of the secular East German system, it might be fruitful to read their reactions against the effects of consumer capitalism in devotional terms. If we accept a reading of consumer capitalism as religion (Baudrillard, 1993), and guides as priests mediating our symbolic deaths and entry into paradise, then the students' rejection of the guides can be interpreted as a rejection of these ascriptions of consumption-as-religion. Metaphorically 'God as the market' should not be the arbiter for our entry into paradise; we need to rely on ourselves for salvation and happiness. In a sense, it becomes a re-working of secular debates in German society and an attestation to the variability not only of religious experience in Germany, but also to divergent views about devotion to the market.

Reflexivity and Parody

A further aspect to this divisive use of guidebooks came in the form of parody and appropriation. To pick up on a theme we briefly introduced earlier in this chapter, Anna and Katrin's use of the guides was at times distinctly critical, part of the idea of the post-tourist. Our conversation with these two particular tourists was punctuated regularly by parodic citations from the guides they were using to travel with, and equally parodic descriptions of those using other guides. Reflexivity and parody can be used by tourists as a way of maintaining a sense of their own différence in the face of so many others, from their own country, who were doing the same as they were, in similar ways to the responses of hosts in areas of mass tourism (Pi-Sunyer, 1978). The guides were read critically for their absences and for their flaws:

> Mückencreme – das steht in keinem Reiseführer. Hast du deinen Regenmantel dabei? Ja es steht im Führer (laugh).

[Midgie repellent – that isn't in any of the travel guides. Have you got your raincoat with you? That is in the travel guides!]

In this example, the guidebooks had failed to provide an ontological security blanket – they did not protect them from the complexities and realities of their being in the destination. Perhaps as a consequence of this, the role of the guide seemed to lessen to some extent as a critical distance was forced through such situations. The security was an anticipated need for the tourist, and sometimes it could not shield them from the material realities of being tourist. The ways in which the guide failed in its job as security blanket in fact became both topic and resource of conversation as demonstrated above.

Guidebooks offer limited protection and are no substitute for experiential knowledge. There is an emphasis on the need for self-reliance and self-discovery – in a way we see an interesting shift here from the guide as a way of institutionalising tourist knowledge and behaviour, to the guide as a discarded artefact, rendered obsolete by the everyday process of 'getting on' with tourist life. Oral culture, in this aspect of the tourist game, aims to trump written culture. For some tourists, frequent visits to the island meant that a guide was no longer required and if anything, it got in the way of seeing the 'real life' of the island.

Concluding Thoughts

This chapter has sought to work with the phenomena of tourist possessions, of things and of feelings. In particular, we have argued that tourism, as we have encountered it, involves alternative forms of work or work-in-rest. It involves work with things and work with feelings, and work with the phenomena and perceptions that claim us through touristic encounter. It involves a labour to reinvent and reroutinise everyday life. It entails more than commodified, consumer relations and more than luxuriating in leisure – though it is embedded in both of those modes of being. In particular we have been concerned to consider the objects and languages involved in the work of intercultural exchange in tourism. Again, we have not seen material life and linguistic life as separate entities. Rather they are embedded and intertwined in the oral and written cultures and narratives of tourism, and in their sensibilities.

Most interestingly, and returning to an earlier point from Chapter 6, our focus on travel guides as preparation for a small, symbolic death, as an assurance against the fickleness of fate, raises the question of the guides' use as an ontological security blanket. It returns us once more to the almost fetishistic contemporary preoccupation with preparing for travel, of which the expanding travel guide industry seems to speak. As a modern cultural artefact, one might simply suggest that the importance of preparation speaks of a modern preoccupation with order and the provision of security through the artefacts and totems of civil society. It speaks of what might broadly be considered a Modernist belief in the purportedly sovereign individual's capacity to control the future and to stride fearlessly into the depths of the dark unknown. As we suggested in Part 2, it is within this context of Modernity's role as either human emancipator, iron cage or disciplinary matrix that much of the questioning about the role and function of travel guides has taken place.

Such an interpretation overlooks the devotional aspect of guides as a form of apodemic literature and the centrality of life-death relations in the emergence of modern culture. Baudrillard (1993: 147) points out: 'our whole culture is just one huge effort to dissociate life and death, to ward off the ambivalence in the interests of life as value [. . .]. The elimination of death is our phantasm, and ramifies in every direction: for religion, the afterlife and immortality; for science, truth; and for economics, productivity and accumulation'. The travel guide industry might then be said to be capitalising on life, by rectifying, by bringing 'order' to any ambivalence of life and death.

So, it would seem, travel guides provide us with life. Life is thereby, literally and metaphorically, the lifeblood of the kind of capitalist accumulation promulgated by the travel publishing industry. It banks on a metaphysics of anguish, and offers a valorisation of life. The experience of tourism from which our focus on guidebook use has emanated shows how the material practices of life – that is living – move this double structure of life and death, which had been surfaced during the ritual moments of preparation for departure and preparation for the next day's touring, into the background once more. Death is eliminated through the forgetfulness of living and the usefulness of the travel guide fades. The privileging of oral narratives move the tourist out of the written separation that the literature and guides enable and into more immediate aspects of being present to place. They enable a change, cultural, social and personal. Being guided orally gives higher narrative capital than reliance on a mere written guide.

Equally, the use of the travel guide lessens when tourists realise that it cannot protect them from the challenges and dangers of life, where it fails to give warning and protection against the unexpected and the unprepared for. Here then lies a 'double blow' in the thanatological function of the guidebook as it is used in everyday tourist life – its usefulness is outlived and gone beyond when faced with the 'getting-on' of life; and by the same token, its uselessness is also highlighted by the unexpected disorder of the real tourist world.

In sum, this chapter has broadly followed Lury's (1996) dual lines of analysis to describe and interpret the different relationships that tourists have with a number of different kinds of exchange – material and linguistic – involved in the process of reinventing bodily habits as described in the previous chapter. From this perspective, we have shown the manner in which practices of exchange, story-telling and object-relations intertwined in different contexts and in different ways to instantiate central modes for the 'doing' of tourism. Whether exemplified in the use of travel guides, or the encounters on landing stages, exchange and oral narrative formed the lifeblood of our travelling cultures. Occurring in times of spontaneous communitas, for good and ill, such a variety of creative exchanges were generative of and generated by intercultural communication.

Chapter 9
Changing Spaces

In the previous chapters, we explored the problem of the unfamiliar body and used this idea as the basis for arguing that tourism can be interpreted as being about the reinvention of the habitual. We emphasised the multiple materialities of tourism (material needs, practices and objects) and we saw a diversity and creativity of object-relations from which emergent structures of exchange were developed. These were all key ways of reinventing the habits of the body in an unfamiliar space. In Chapter 8 we developed our exploration of this material base of tourism by focusing on a largely undisclosed aspect in tourism research. That is the fact that the material life of exchange, part of a broad tourist talk as we called it, grew out of and took place in language and in narrative. We looked first at tourist narratives and their social, performative, formative and temporal functions. We then turned to look at the uses of travel guides within the context of our research. Exchange and narrative were, we argued, the lifeblood of our travelling cultures.

Driving the insights of the previous chapters was the question of what it was that seemed to facilitate or to hinder the flourishing of intercultural communication. Our answer, that it lay in the urge and enactment of exchange and narrative was, in the main, spatially confined to our analysis of the youth hostels in which we stayed. It was in the spaces of the hostel that intercultural communication most often flourished.

In this chapter we dwell to a greater extent on the question of space and its relation to tourist intercultural communication. It was certainly not the case that intercultural communication blossomed in all the spaces which we inhabited whether as hostellers, guests, visitors or as part of a guided tour. What we found was something akin to a spatial distribution of intercultural encounters on the Isle of Skye. Some spaces furnished the conditions within which intercultural communication was materially possible. In others, intercultural communication was systematically frustrated. Taking the youth hostel as a kind of paradigmatic spatial type for intercultural communication, this chapter compares and contrasts the sociospatial conditions of other places with that of the youth hostels from the previous chapters. Here we draw upon ethnographic slices from three other spaces (a guided tour of the Peaty

126

distillery; an overnight stay in the Flora Bay Hotel; stays in bed and breakfasts) to do this comparative spatial work.

At the heart of the comparative work in this chapter is the question of the conditions of possibility for intercultural communication and the manner in which these are mediated and refracted by sociospatial concerns. Key to this concern are the questions of commodification, consumption and control, which we pointed to in Chapter 2. We might take something of a cue from the following passage by David Chaney (1993: 164–165), worth quoting at some length here:

> As tourists we are above all else performers in our own dramas on the stages the industry has provided. Although clearly constructed landscapes of theme parks and pleasure grounds are exceptional, in practice all tourist locales can be seen to involve degrees of staging. There is in all of them a management and clear articulation of cultural identity oriented towards what Urry (1990) has called 'the tourist gaze'. To say this does not imply that tourism is an unchanging practice, we should rather see the history of tourism, or the development of popular holidays, as encapsulating the theme of transition from crowd to audience. (...) it is in the practice of gazing that tourism is staged. A tourist is a distinctive type of stranger, collaborating in the spectacle that is being performed. The dramatic representations are therefore staged for a potential audience. The managers (...) of tourist sites draw on narrative formulae much as the producers of other media of mass entertainment such as film and television. They cannot, however, fully control how they will be seen and appropriated.

The notion of a transition from a crowd to an audience implies, for us at least, a process of commodification through which we become a set of identifiable consumers with values, needs etc., that can be satisfied through money-based exchange, with some kind of tourist experience. Central here, as Chaney articulates in the final sentence above, is the question of control and the extent to which producers determine the meaning, behaviour and experience of their audience or whether the audience has the capacity to appropriate or even change the intended message of the producers. We suggest that the tensions between these structures and agents implicated in such questions are locally enacted, embodied and take form in the interactive relations between producers and consumers. We might understand these as reflecting the tensions, mentioned earlier, that Turner elaborates between liminal and liminoid forms; or, as Schechner sees it, between efficacy and entertainment.

Using this local perspective we ask about the differences in our, and other tourists' experiences of intercultural communication across and within these spaces. In short, do heavily commodified spaces succeed in promoting intercultural communication? Or do they hinder it? If so, how? What kinds of social relations does it involve? We will argue that these different relations have divergent effects on the prospects for intercultural communication.

Conceptually, two key themes emerge from our analysis. First, we argue that the relations between sociospatial forms like a hotel or a distillery and intercultural communication are conducted through a particular political economy of the body. Using Foucault (1991), we try to explore the connections between the intercultural bodies of tourism and the manner in which commodified spaces provide panoptic regimes that encourage tourists to engage in self-surveillance of their bodily conduct. In the process, they usurp the need to engage in any sustained form of intercultural communication.

As such, we examine how tourist bodies are subjugated and subjected to the regimes of capital articulated in these spaces, and point to the governmentality of the tourist body through the movement of power relations. This is achieved through the close control and supervision of bodily movements in time and space. Adding some insights from de Certeau (1984), secondly, we build upon this Foucauldian interpretation through greater reflection on spatiotemporal dimensions under capital through his notions of 'strategies' and 'tactics'. Now, to the bodies in space.

The Peaty Distillery

The Peaty Distiller is a producer of a well known brand of whisky and is owned by a multinational. It is a frequent destination for tourist buses and others on the whisky trail. The whisky trail, a particular set of routes allowing tourists to Scotland to visit all its distilleries, is very much a key stage for the production of Scotland the 'holiday destination'. The day that we went to the distillery, we arrived before the inevitable appearance of a multitude of international tourists. Consequently, rather than being shown around with a linguistically and culturally homogeneous group of tourists, we were part of a set of visitors who had made it to the distillery before the morning rush.

Our visit to the distillery was organised into very distinct spatial and temporal tranches. We found out that this seemingly well considered form of organisation was the result of a recent refurbishment to the distillery, which turned it from a working location producing whisky to a working tourist attraction. On entering, we came into a foyer with pictures and photos hanging on the walls that spoke of the distillery's past. There was also a small number, perhaps two or three, of locked display cabinets in the centre of this relatively small space containing the tools and associated material objects of whisky production. There was, of course, a cash register area where a cashier took money from us in exchange for a guided tour of the distillery. Protected by a counter on which lay various maps of the place and other historical information, the cashier was also responsible for providing all of us waiting for the next tour (there was one every hour) with a nip of whisky. Full bottles were naturally available for purchase in the exit shop after the tour.

Whilst enjoying the qualities of the Peaty product in the 20 minutes before our tour, we took time to look at the black and white photos and pictures of old men working with whisky casks and other such things. Here we were presented with 'tradition', which in this particular case constituted a small slice of the patriarchal organisation of the public sphere of work. This was a glimpse back to an early time where mass-manufacturing processes enabled a greater production of whisky, and where this work was carried out by men. We immediately noticed material objects contained within the locked cabinets, those once wielded by the hands, the prosthetics, almost, of physical labour. In a distillery where whisky is no longer made, the possibly unpalatable truths of the workings of multinational companies are concealed through appeals to tradition.

Amidst this 'production' of tradition, we were also struck by the large number of 'quality certificates' that graced the foyer. They were mainly located behind the cashier's desk. There were prizes for the distillery as a top tourist attraction and other awards sponsored by the Peaty Distillery (or its owner's multinational marketing arm at least) for other visitor attractions on Skye. It seemed that Peaty was some kind of benevolent sponsor of a set of awards for other attractions on the island which could be voted for by tourists. On the desk there were forms for visitors to vote for their favourite tourist experience on Skye, with a promise that they would be put into a draw to win a bottle of Peaty if they entered. Perhaps, we wondered, this was an instance of expectation management. Surely this distillery must know what a good visitor attraction is given that it sponsors awards for this very thing! We were excited.

Here we had a preface to our guided tour. On the one hand, we had a preface that told us to expect a historical and technical insight into the male world of whisky production on Skye. There was a clear articulation of a set of heroic and proud values of the legacy of whisky production on the island. On the other hand, we had a more contemporary commercial preface, which simultaneously positioned the distillery as experts on quality tourist attractions and us as consumers judging the experiential quality of these attractions. We had the juxtapositioning of the highly traditional and the hypermodern coexisting in one space.

In this moment we moved from being a crowd to an audience, one positioned with purpose and intention by the distillery owners. However this reaching out to us with historical insights, Scottish hospitality and a consumer 'voice', failed to materialise in the practice of the main tour. For what transpired through the tour was not the warmth of a group of culturally and linguistically mixed tourists sharing time with the insights of a tour guide. Rather we were on the end of the delivery of a highly scripted and controlled performance on the part of our tourist guide that offered nothing in the way of spontaneity, surprise and, crucially, for our lines of questioning, intercultural engagement and exchange.

We were part of a group of around 20 tourists. The group consisted of different nationalities and language groups. Having had our wee nips of whisky and consumed texts about whisky heritage and quality consumption, our guide came to meet us and to introduce herself and the tour. She was a young student at that time studying Geography and Gaelic at a Scottish university. This was a summer job for her. Having had a material cultural preface from our interactions with the objects in the waiting area, as well as a welcome and preface to our visit from the tour guide, we then moved through into the first area of the distillery.

The guide stopped. Our group caught up and formed an untidy circle around her as she began to tell us about the initial stages of whisky production. The door between this first holding area and the foyer had been closed, and we were concealed in this small space listening to the narrative and looking at the various artefacts and objects to which her story referred. The tour was in English, the only language available for guided visits through the distillery. As there was no opportunity to wander through the distillery at leisure (you had to be part of a guided tour), this meant you were forced into hearing the delivery of a particular story about the distillery in English. There were, however, what might be described as 'translation cards' in a few select languages. These rather basic cards simply explained technical details relating to the objects around us and their part in the whisky production process. After 8–10 minutes of her speaking, she asked if we had any questions. No one said anything so we moved into the next part of the distillery.

The next section was spatially distinct from the last. The tour guide led the way, stopped, waited for the group to form into a circle to listen to her and then continued with the technical details of the next stage in the production process. Again after 8–10 minutes of effortlessly delivered technical and heritage details, we were asked if we had any questions. This time, the patriarch of an Italian family asked a couple of questions in English merely to confirm, have explained or repeat certain points that the tour guide had made. In response to these questions, she merely repeated her script, with very little variation from the first time around.

This routine of moving from one delineated section of the production process to another, forming into a circle, having 8–10 minutes of a script delivered at us, being asked if there are any questions, hearing very few, and a repetition of the script to those that dared to ask, continued in the five other 'sections' of the guided tour. It became something of an emergent routine, a way of coming to know how to behave and comport ourselves in these new surroundings. Whilst it was interesting to hear about whisky heritage, the large amount of technical detail on the tour bored us. We wondered whether you either needed to be a 'technical' type to take any intrinsic interest in the description of a whisky production process, or a predictable tourist feigning interest in the deathly dull detail of it all for the sake of being able to say that you had 'been to' or 'done' a distillery. Neither of us can recall a single detail from the guided script in retrospect.

The guide's role in strategically managing not only the narrative she delivered, but also our movements around the distillery was crucial. She played a pivotal role in reproducing and keeping in place a particular story about whisky production on the Isle of Skye and Peaty's part in it. On one level, this management took temporal form in terms of the carefully controlled amount of time we spent at each section of the distillery – no more than 8–10 minutes per story, and the obligatory didactic request for questions which we did not avail ourselves of. On another level, there was spatial control in terms of the supervision of the movement of our bodies through the distillery – making sure that we did not spend too long in any section, waiting until the group had circled around her before delivering the script, making sure we were quiet before she started. Perhaps most interestingly though was the handling of the questions. One way she ensured that her script, and the dominant reading of the distillery would not be disturbed, was simply to repeat the lines she had been given. Perhaps this was understandable. She was a tour guide, not a distiller.

One question asked was about the political economy of the whisky industry and the effect of multinational ownership on local communities. It served to highlight the partiality of the script she was delivering. We knew that working in the distilleries of Skye used to provide many men with jobs. Industrial change and ownership patterns (mainly in the form of rationalisation and technical advance) affected this profoundly and led to workforces being cut. The history of this distillery was no different. Their multinational owners saw over a rationalisation of local and national distillery workforces. Of course, this political economy did not feature in the narrative that was told about whisky heritage.

When asked this question, this was the only time that 'ums' and 'errs' appeared in the tour guide's speech. Improvisation was required, and none was forthcoming. She did not have fluent answers to these questions. This seemed strange to us, though, given that she was a native of Skye and had made much about her personal cultural heritage at the start of the tour. Surely she would have known about this. She smiled though, and delivered her answer in the sweet lilt of her Scottish accent. Having dealt with this 'dissent' to the official story of the distillery, she managed to steer things back on course for the remainder of the tour.

What stands out above is the disciplinary regime that existed in the distillery. That is the controlled movements of the tourist body in time and space. Following Foucault, the body of the tourist was subject to the constant supervision of the tour guide as part of a wider apparatus that partitioned space and movement. The script the tour guide delivered was also produced through diffuse regimes of control. This was a strategic space, in de Certeau's notion of the word, to which we return later. In other words, here is a space where place wins out over time. Through the production of heritage, of history, first of all, time has been commodified and exposed to the linear logic of the interests of capital. Second, through the close temporal control of the tour, the guide makes sure that we take more or less a schooled hour, give or take a couple of minutes, such that she can begin the next tour; so as not to interrupt the schedule of visits which divided the days into the schooled hours for lessons.

A further point of note is the pivotal role of the tour guide herself and in particular her role in delivering and negotiating the tour script. Here she was reproducing and keeping in place a particular and highly partial story about the distillery. Was it the case that she was delivering a story that was simply 'written into' the tourist mind? Were we all duped into believing what we heard? Were we not as consumers, to some extent at least, involved in appropriating the meaning of this story? We would argue that dominant readings associated with a particular act of consumption do not just inscribe themselves onto the tourist experience. Although our bodily conduct was highly circumscribed, even schooled, in the distillery, it was so because of our active involvement as consumers in this process. It was part of a system of self-surveillance associated with the rationalisation of this space.

Centrally this system of surveillance was enacted in the relationship between the tour guide (in this case a 'producer' – the agent of capital) and the consumer. In this regard the inter-relation between us was active, engaged and the site of the control of our bodily conduct. We could always have run around the distillery, shouted out obscenities and told the guide she was talking nonsense. 'What about the workers?' we might have said. In terms of the story she told though, the potential for undermining the dominant script was possible in the question time, as suggested earlier. The majority of the tourists were predisposed to accept all of the story they were told, mainly because it was a technical one. Science's script is often unintelligible in lay terms and beyond criticism. However, the acceptance of this story indexed a particular set of forces of power involved in consumers taking upon and accepting the readings of the distillery provided by the tour guide.

The guide was 'keeping in place' the story. This present continuous of 'keeping' is important, as it suggests that reception and acceptance of the dominant reading is an achievement, a product of the subtle negotiation between producer and consumer. This process may unfold in more or less ambiguous or contested ways, but the important thing is that it has to be worked at. It can never simply be accepted. As Foucault (1978) reminds us, power does not write itself. Power is not something that one has or possesses. Power is a part of relationships, not something that is foisted upon us. Power provides the conditions for its own formation, but it requires an agent to make it work. It also furnishes the conditions for resistance and there is indeed a contiguity between relations of power and resistance. As such we might read the role of the tour guides as one of policing the contiguous lines between power and resistance inherent in the inter-relationship between producer and consumer. This was a subtle, embodied and precarious process in which the party line was produced and kept in place.

The consequences for intercultural communication were, perhaps not surprisingly, stark. There was no discursive engagement between the groups and the dialogue was between the tour guide and the people that asked questions. There was no interchange between the tourists during moments of movement. We were fee-paying strangers. The conditions, those of surveillance and strategic management, were far from propitious for intercultural communication. In fact, there was a systemic frustration of reciprocal dialogue (Scott, 1990) and interaction between the tourist groups. This had more in common with West End theatre than with communitas. It was liminoid, not liminal (Turner, 1982). These highly scripted and managed performances offered us little in the way of intercultural engagement. It was as if we were being taken through a set of well managed performances, which became impenetrable to the citational practices of the moment. At best, it involved a scripted set of cultural interactions between the tour leader and the few tourists that asked questions; at worst it was simply a multilingual environment in the sense that it contained merely translation cards and we were able to identify alterity by different language communities. This impenetrability was of course held in place, softened and sweetened through its embodiment in the tourist guide – those sweet eyes and soft local accent. The reflexive monitoring and conduct of the self created distance, which individualised rather than socialised.

Having said this, our visit to the Peaty Distillery was not completely devoid of intercultural interest. In any case, the wee dram had lightened our spirits from the start. In order to get out of the distillery, you had to go through the obligatory exit shop and its visitors' comments book. Here things actually became more exciting! For one, several of us seemed to stop at the visitors' book to look at what had been written rather than to write something ourselves. Apart from the usual scripted pleasantries, we also found these three comments:

Wow! Sauteuer: das Zeug hier aber gut (Germany)
Why don't you arrange guides in other languages?
Slainte (Essen)

The first complains that the distillery is exorbitantly expensive, but that the whisky is good. The second points to the lack of guided tours in languages other than English. And the third uses the Gaelic term for 'Cheers' and was written by a visitor from Essen in Germany. We were quite taken by the first two negative comments, as bland and compliant ones are usually the order of the day for these kinds of books.

Our interest was piqued, to our surprise, in the exit shop. Having lurked around the comments book for a little while, we were the last to enter the exit shop. When we did, we witnessed a vastly different scene to the one from the guided tour. Here, people were much more animated, talked to each other, exchanged furtive eye contact when deciding which particular whisky to buy. There was more motion, talking and nonobservation of physical boundaries here. People moved about, smiled and laughed with each other over the hairy haggis and the furry Loch Ness monster. They bumped into each other and apologised. Sometimes they exchanged a few words and some wry smiles. Perhaps we all knew we were moved there to buy some cheap, and some rather expensive tat. But it did not seem to matter, as most of us engaged enthusiastically in this exit shop. Perhaps this was because we were all so bored after the tour, that any light amusement would have been welcomed, rather like playtime after a dull lesson. Perhaps also we were more familiar with the routines of an exit shop and we could all just get on with it. Perhaps too there was less control of our bodies and minds here. Whatever the case, the exit shop provided us with a chink of colour in the grey and drab surroundings of a distillery tour.

The Flora Bay Hotel

The strategic management of time and space, and the concomitant regulation of the tourist body was also clearly illustrated by our experiences in the Flora Bay Hotel. Here too was a space characterised by the systematic frustration of dialogue and reciprocal action (Scott, 1990). But in this particular place, the conditions and dynamics under which this frustration of intercultural communication took place were different. The difference lay in the fact that the Flora Bay is one of Skye's premier hotels and as such, the tourists there would certainly occupy privileged socioeconomic positions. This was a place that attracted a group of rich tourists from different cultures. This meant for different dynamics than the guided tour of the distillery, but ultimately the same result for intercultural relations. The dynamics of intercultural isolation were founded upon particular forms of symbolic consumption based in particular socioeconomic types and expressed through the regulation of the tourist body in this political economy.

As with the last case, however, the kind of cultural dislocations did not just happen as some kind of obvious social fact. This was the product of a complex set of spatial and temporal arrangements, reflexive monitoring and bodily conduct and particular kinds of civilising process associated with dining. What we might, following Foucault call 'tourist subjectivities', that emerged in this situation were underpinned by a marketing-driven discourse based in the provision of conditions for particular kinds of status-based symbolic consumption which had individuating effects. It involved a mise en scène of a set of control structures, which mitigated against spontaneous social action. And these control structures lay in the complex and embodied relations between staff and customers at the Flora Bay.

We were very much looking forward to our night of luxury in the exclusive Flora Bay Hotel. Having spent the previous night in a very uncomfortable youth hostel, a comfy bed and a clean toilet seemed like heaven. Both its promotional materials and its word-of-mouth reputation suggested that this was a quality hotel, offering high standards of food, accommodation and service and a relaxing break. It was also a tourist space that we knew would differ considerably from the youth hostel. The Flora Bay was spacious, peaceful, off the main road and 'away' from the crowd, its exclusivity enhanced by the small, tree-laden country road, which disconnected the hotel from the busy road nearby.

When we arrived, we decided to sit outside and began to observe what kind of place this was, who was there, and as ever, to have a complimentary cup of tea. Again our observations involved looking for signs of interculture. We looked for T-reg hire cars and at registration plates. We looked at the inscription of national cultures on clothes. We listened for languages as markers of difference. It became clear that this too was potentially a good place to observe and experience intercultural communication as we saw signs of Americans, Italians, Germans and French in and around the hotel.

We decided to have our tea at a table area immediately in front of the hotel reception. We ordered and waited for 20 minutes or so. A waitress immaculately dressed in black and white served our tea on a silver tray with a whole manner of different crockery and condiments made from the same fine china. Despite trying to engage the woman that brought our tea with questions about the hotel, the 'season', the nationalities, as we had done with others that we had encountered, her responses were short and polite, but hardly forthcoming. Whilst having tea, we became aware of two French teenage boys and their Scottish hosts (a couple) having lunch at the table next to us. They spoke French to each other, although the older boy was speaking a mixture of English and French. He asked for some vinaigrette for the salad that accompanied his steak. At this point, the Scottish male host retorted, in French:

Les gens en Ecosse ne connaissent [sic] pas faire une vinaigrette. [People in Scotland do not know how to make a vinaigrette.]

Shortly after this, the younger French boy got some mustard caught at the edge of his lips. His female host took her serviette and wiped her mouth, gesturing the young boy to do the same. He followed her gestures.

In the unfolding of this interaction, we suggest that these well heeled Scottish hosts were playing with the different symbolic and material culture of this space, using it to position themselves as hosts to their guests. Following Bourdieu (1984), the hosts were clearly displaying their knowledge of a particular set of social codes around the preparation of a French salad and keeping one's mouth clean after eating. The comment about vinaigrette might be read as a particular expression of distinction. It very much fits in to the emerging food culture of contemporary Britain, encoded in the proliferation of TV cookery programmes and glossy Sunday supplements. Talking about vinaigrette is a move to distinguish oneself from what we might assume to be its invisible supplement – the unsophisticated Scottish diet. It is also an attempt to identify themselves with the French boys whose palate, as French, is of course presumed to be more sophisticated.

Following this set of exchanges, we can see that the Flora Bay, as a place, has a particular symbolic order, around the cultivation of upmarket Scottish hospitality, as much as it has a particular form of material production (particular kinds of furniture, imposing nature of the buildings, service). We saw how these locals seemed to be using the setting of the Flora Bay for their work of identification but also distinction.

We moved from our outside table into the building to check in and check out its facilities. Upon entrance, we were immediately hit by the raft of quality certificates on the wall, the visitors' book with the usual set of mimetic structures of guest appreciation and the quiet, efficient, self-effacing welcome offered by the hotel. The man that checked us in made very little fuss, taking our details with the minimum of verbal and bodily cues. He talked quietly and we wondered, just for a second, whether we too had to go around the hotel whispering to each other. We agreed to meet later in the lounge of the hotel for a gin and tonic just before dinner.

The lounge was lavish. It had a large fireplace and mantelpiece, art on the wall, extra furniture. There were a number of couches spaced out in ways which divided the room into particular places – their boundaries demarcated spaces for interaction. What happened was that those groups or couples who were staying in the hotel together occupied particular couches and therefore inhabited an exclusive space which stunted any possibility of interaction. Although there were occasional glances, smiles as we brushed past each other, no engagement was pursued. We teased our gins and tonic whilst watching an American-sounding couple sitting on one couch, and a large non-English-speaking family at another.

Having finished their drinks, they moved into the dining room leaving us in the lounge to chat.

On moving into the dining room, we found yet another opulent space with expensive looking furniture and a flourish of tasteful tartan. As well as us, the North American couple and the non-English-speaking family, there were two other tables occupied for dinner. Rather than being put close to each other, we were all spread out in the dining area. The distances between the tables were already significant and precluded any chance of being distracted by other dinner conversations. The conscious spacing by the waitress of the guests served to enhance the likelihood that 'privacy' would be gained during dinner. The couple, the group and us were all given menus in turn and, following some brief negotiations with each other, we all ordered. The waitress taking the orders was an older woman dressed in the same black and white attire of the tea-server from earlier. She conversed with us very little, simply asking for our food and drink requirements politely and again, we noticed, almost in a whispering fashion. We had that feeling that there must be someone important that should not be disturbed by anything like normal-level conversation.

We were curiously bored. Others were too. We noticed the North American husband reading a book at the dinner table ignoring his wife and seeming irked at his wife's continuous attempts to make conversation. A family chatted together quietly and intimately. We felt isolated sitting at our little table. And it seemed that this was meant to be so. Our communal time was closely managed by the waitress who performed all the ostensibly good structures of quality consumption. The performance of hospitality was safe, clean and expensive. She was there to ensure that all bodily conduct was regulated, and in a whispering tone too.

Having tea outside, sitting in the lounge, having dinner, having breakfast – these four scenes were characterised by a lack of interaction and dialogue between all the tourist groups we observed. There was some acknowledgement but that was as far as it went. The reason for this, perhaps, lies in the symbolic consumption in this highly marketised space. Tourists here were paying for privatised space in which the work of the body, such as that discussed earlier, is either conducted behind the closed doors of the en-suite bedroom or highly regulated by the material space of the hotel or the actions of the staff. There was no compulsion to interact with others, and perhaps there was even a desire to be left alone. But this is not something that just happens.

The occlusion of social interaction, and thereby intercultural communication from this space, was an achievement of its relational properties.

It was carried out through the highly structured regulation of tourist bodies via partitions of time and space. The consumption of upmarket, privatised Scottish hospitality was an altar for securing particular kinds of subjectivity. There was a clear object of consumption, a commodified and well rehearsed hospitality that located desire and power in the material culture of the hotel and tied its inhabitants materially and symbolically to this locus. Securing the status as an upmarket tourist involved highly individuated engagements with this object, which turned tourists away from engagement with each other. The unfolding of acts of consumption had a dislocating effect on social relations. It was a place of real intercultural poverty. The potential charisma of a Scottish welcome had been well and truly routinised in the bodily scripts of the producers.

What we find in these two examples is a distillery tour that has the feel of a Victorian school classroom, or even a prison, and a boring social experience in a hotel. At one level, it should hardly be surprising that the guided distillery tour and the hotel were places of systemically frustrated intercultural communication. They were hardly set out to facilitate this. These were privatised spaces in which tourist bodies were controlled through a series of surveillance mechanisms, importantly embodied in staff–customer interactions, where spatial relations trumped temporal ones. The youth hostels dwelled upon in the previous chapters however, were public spaces where social integration was, for the vast part, a necessity for getting on and dealing with the reroutinisation of the body. As this was a public act, the other was necessarily involved in understanding the emergent rules for negotiations of the unknown, the uncertain and the spontaneous.

What we are at pains to point out here is that this cultural dislocation is not an inevitability or an easy to accept social fact. These more or less individuating or socially integrating experiences were outcomes of social and discursive interventions at the level of the embodied social relations of capital. They are certainly conditioned by their position in systemic relations of production, but they do not write their own histories. For this, agents are needed. This will raise ethical questions, but for now, let it be said that whether or not these are desirable social situations for tourists depends of course on the tourists themselves. For us, there were delights and discomforts in each.

Perhaps most interestingly though is the case of bed and breakfasts, which occupy a curious middle ground between the public and private spaces of the youth hostel and the Flora Bay (on some kind of imagined accommodatory continuum). Bed and breakfasts are of course highly diverse types of space. On the one hand you have those that try to enact the structures of consumer culture. In Scotland, and we noticed this in a couple of bed and breakfasts, the need to get accreditation from the Scottish Tourist Board (currently called Visit Scotland) involves assuring that certain things are provided for guests. There must, perhaps, be coffee- and tea-making facilities in each room; perhaps the capacity to pay by credit card. On the other hand, it must also be remembered that bed and breakfasts are often, although certainly not always, owned by private individuals or families who often live in the same space. As such, hospitality is less clearly defined in the terms of Visit Scotland and more often involves a simple opening up of the family home to others.

B&Bs are tricky. You never quite know what to do and how to comport yourself. Perhaps this is because they are an uneasy mix of the familiar and the strange. For us, this was most often felt in the breakfast room of the B&B, for here was a space in which pregnant pauses and interminable silences reigned, and a kind of shared sense of embarrassment was almost palpable. The familiarity stems from being literally in someone's front room. It has all the signs of home – photos, a lived-in look. But the strangeness comes when finding out the rules that make this space work. Can you just go and help yourself to cereal? Are you restricted to two rashers of bacon? You never quite know whether you can really treat this place like home, as the owner is there to confuse the situation and ensure that you confer with their routine. There is also an uncomfortable silence in breakfast rooms. You often feel that you cannot say anything because no one else is and if you did, all would be able to hear your tittle-tattle. So instead we all sit there sharing our embarrassment and willing for those sausages to arrive in order to give us a purpose or at least a source for conversation.

Strategic Spaces and Temporal Tactics

Drawing together our reflections on all the places we have looked at (youth hostels, a distillery, B&Bs) we might say that unpacking the tourist bag and the engagement in intercultural communication was mediated by a plurality of different structures for exchange. We might call these different structures for exchange 'alternative economies'. These entwined, connected, disconnected from each other in a myriad of different material and cultural exchanges. Here we have market- and consumer-based exchanges in the distillery and the hotel; oppositional, 'post-tourist' economies in youth hostels; there are also multiple sets of exchanges between visitors, part of an informal tourist economy for the swapping of stories, advice and cynicism. These are tacit and emergent knowledges shared 'secretly' between tourists, not really instantiating so much a logic of the market, but often creating some kind of communitas. And then there are the emergent and unexpected material exchanges of the youth hostel centred around the rituals of everyday life and the reinvention of bodily habits.

Part of touring around the island involved weaving in and out of these different spatial economies and experiencing Skye within their constituent systems of exchange. Each of these structures exhibited different forms in the intellectual and material sense of the word. In ideational terms, these economies worked with different logics of exchange, ranging from those based on already structured, rule-bound market-exchange to those based on an emerging sense of communitas. These differed in terms of the kinds of values they produced, structures, rituals, practices and poetics and their emergent sense of the intercultural. They were served by different material needs and exhibited different material cultures.

In short, these different but coexistent structures of exchange endowed our experiences as tourists with different form as they transformed the cultural space for relationships. Our journey on Skye then was characterised by shifting relationships in shifting cultures. This transformation of the cultural space of relationships is vital in accounting for variegated forms of intercultural communication we encountered on the Isle of Skye. The story that we might tell then is a changing story of intercultural exchange.

The spatial comparisons in this chapter suggest that the systemic nature of tourism cannot be ignored in accounting for intercultural tourist interaction. Relations between production and consumption, and the mediation of tourist space and time by the interests of capital and its various structures and technologies, clearly impacted on the fate of intercultural communication in a tourist context. Here there is a set of intimate connections between political economy, the tourist body, space and time to be unravelled.

It is here that Foucault and de Certeau can be instructive in explanatory terms. As noted earlier, de Certeau (1984) for instance argues for a distinction between what he terms 'strategies' and 'tactics'. Strategies, he explains, are a calculation of power relationships which involve the 'postulat(ion) of a *place* that can be delimited as its *own* and serve as the base from which relations with an *exteriority* of targets or threats (customers or competitors, enemies, the country surrounding the city, objectives and objects of research, etc.) can be managed' (de Certeau,

1984: 36, italics in the original). Strategies are actions that, in being exercised in particular places, involve the triumph of space over time.

The Flora Bay Hotel can be viewed as a strategy of capital: the elaboration of a particular place involving movements of power between producers and consumers. Through its elaboration, the time associated with Scottish history and hospitality has been translated into the commercial space of the hotel. Time here becomes continuity: history read within the present interests of capital and translated into the controls and routines of the consumption acts associated with the hotel. Importantly, according to de Certeau, one of the key effects of strategies is that actors come to master places through sight:

> The division of space makes possible a panoptic practice proceeding from a place whence the eye can transform foreign forces into objects that can be observed and measured, and thus control and 'include' them within its scope of visions. (de Certeau, 1984: 36)

In terms of the hotel then, it is its visual culture, encoded not only in object-relations but also in staff–customer interactions, whence panoptic practice and its constituent forms of self-surveillance take place. Here is the space of the docile bodies, of frustrated intercultural relations. The putting into place of mirrors for consumption serves to eroticise the relationship between the subject (the tourist) and the object (upmarket Scottish hospitality). Through the commodity relations indexed in the articulation of this hotel, we are compelled to invest in the regulatory structures of the tourist body which individuate the consumption experience and frustrate our time with unknown others. The hotel becomes a site for the eroticisation of the self through commodity relations, pursued via the unchanging and continuous trajectory of capital.

By contrast, the youth hostels, as places of normative communitas, are anything but strategies. These might better be described by appropriating de Certeau's notion of 'tactics'. Tactics are the actions of the powerless, not the powerful as was the case with strategies. They are determined by the absence of a power locus where there is no delimitation of an exteriority which provides it with the condition for autonomy (de Certeau, 1984: 38). Tactics are procedures that lend 'pertinence to time' and 'to the rapidity of movements that change the organisation of space' (p. 38).

Here we recall the confusion and uncertainty associated with learning the routines of the hostel as, in no small part, a question of time, of discontinuity and rupture. We have no easy temporal continuity because of the breaks in our routine and the necessity of renegotiating this with others. Unlike the hotel, there is a less scripted discourse for the creation of an object of consumption, and peoples' attentions are turned to the other rather than to the self for relationships. This lack of an object serves to re-eroticise social relations as a material and symbolic necessity and thus reterritorialises desire into more diffuse times and spaces. It is about the projection of desire onto the other as a real human being rather than the other as a commodified experience.

Going back to those moments of choice, the critical moment, when a smile might have turned into a sentence and into a conversation, the lack of follow-through could be interpreted to be a result of the process of objectification and the rationalisation of the erotic into commodity relations. The stay in the hotel, rather than the human beings that inhabit it, is the other and this would seem to marginalise the perceived need to pursue interaction. In the youth hostels, there is no specific object for consumption in the same way that the hotel has one. This locates hostellers' focus of attention on social relations and intercultural communication in an act that relocates power and desire in the relations of the hostel. In the youth hostel, we have a shifting relationship between space and time. Time becomes the pressing issue – the discontinuities and disruptions to routine. We make time in the youth hostel to create new routines and re-establish continuity.

In sum, this chapter has set out to explore the way that tourist space and time are related to the chances of intercultural communication. What we have demonstrated is the existence of a particular 'spatial distribution' of intercultural communication within which intercultural engagement between tourists flourished in some spaces, whilst being systematically frustrated in others. We suggested that the complex interpenetration of space, time, bodily discipline and commodification was accountable for such divergence.

In the final section of the book, we begin to pack up our travel bags and return home. Our journey is coming to an end.

Chapter 10
The Return to Routine

This chapter takes its leave of the destination and orients itself in the direction of home. Again it works with both the phenomena of real, material baggage of departures and with the metaphors that may be mined analogously from the repacking of experience for the purpose of a return to the familiar. We begin by examining the photographs and souvenirs that are collected as part of tourist activity, and taken back home to be incorporated into life, or given as gifts. We then explore the process of packing and the dirt that accompanies departure and is taken back home for processing. We look at the ritual aspects of washing and their foundational relation to our cultural notions of civilisation. We then juxtapose the dirt with the 'new': the freshly purchased objects; the freshly garnered experiences; the renewed body, and the lines between the two, in order to examine the work of ordering that accompanies a return to normality. And finally, our attention turns to the reintegration of new things and feelings into everyday life, and the change from a predominantly oral to a more written form of tourist narrative.

Collecting Souvenirs

Part of the labour of tourism is, as we argued earlier, that of the future-oriented work of preparing memories and of responding to the claims made by a place, and by people and materials encountered. Such emotional labour in the present to secure future narrative for memories indexes something of a paradox. The taking of photographs and the buying of souvenirs give material, sustainable form to what is ephemeral. It is a way of symbolically distilling present-tense experience into fragments and forms that have a durability beyond tourist time. It makes our tourists 'collectors' and orients them towards future times of memory.

To the north west of Skye there is a memorial to Flora Macdonald, of Bonnie Prince Charlie fame. It bears the following inscription:

'In the family mausoleum at Kilmuir lie interned the remains of the following members of the Kingsburgh family, viz Alexander Macdonald of Kingsburgh, his son Allan, his sons Charles and James, his son John and two daughters. And of Flora Macdonald who died in March 1790 aged 68. 'A name that will be mentioned in history and if courage and fidelity be virtues, mentioned with honour. She is a woman of middle stature, soft features, gentle manners and elegant presence.' So wrote Johnson.'

Such was the inscription on a marble slab erected over this burial place by Colonel John Macdonald F.R.S. of Exeter who died August 16 1831 youngest and most distinguished son of Flora and Allan Macdonald of Kingsburgh. Every fragment of this memorial has been carried away by tourists. In grateful memory these words are now restored by Major Reginald Henry Macdonald O.B.E. (retired). A great, great grandson of Allan and Flora 1955.

There are various ways in which this inscription provides an interesting point of departure. We are interested in the mention of what was absent – the stones taken away by tourists. As described above, stones provided an important source for tourist collections from Skye in the past. Each stone taken away is a symbolic reminder of a place, a mythic past, a moment when meaning was touched, history grasped. It acts as proof, and it is portable in a way that present-tense, oral performances can never be. In other words, these stones come to stand in for experience, to trigger memory and to recall feelings. The stones are not just any old stones; they are animated with meaning and emotion for those who removed them.

Our tourists took away stones and pebbles and pieces of coral from the island's beaches, serving to connect them not so much with national or Romantic myths of people, but with the romance of a beautiful, remote Scottish island. At this memorial, however, they collected photographs, as they did so, commenting on the shocking behaviour of such previous tourist collectors. Here we find tourists following the imaginative trails of earlier tourists, but bringing a different moral sensibility to inanimate objects that helped clarify, and importantly, record, the perceived difference between their own behaviours and those of previous generations of tourists.

Other items served as souvenirs. Children, as we noted earlier, purchased green and tartan soft toy versions of the Loch Ness Monster. Adults, again as noted earlier, took whisky home. Local knowledge was collected, shared, tested, recorded and taken home. In the common room of the hostel at night we found ourselves, almost ironically, as the key informants on all things Scottish and worthy of collection. Gavin wrote out some Gaelic phrases for some of the Germans we had got to know.

Monument to Flora Macdonald

They sent them home on postcards, proudly displaying what they had found.

Photographs

Photography was a constant practice, inserting a break in the flow of tourism to add a layer of reflexivity, a moment for the record. Although we offered to take photographs of our tourists we found that many preferred their pictures to be people free. The stories they would tell back home, using these pictures, were those of wild rugged scenery and 'nature in the raw'. They were stories' that would fit with the public Romantic myth of Scotland's landscape, stories that formed part of the written cultural baggage discussed in Chapter 6 that also blazes imaginative trails into destinations. Although the actual experience of tourism was, as we described it in Chapters 5 and 8, one that privileged oral and ephemeral forms of narrative and culture, the practice of taking photographs and buying souvenirs looks forward to the reincorporation of the images and objects into the pre-existing public and mythic structures and into written, textual forms of narrative.

The Romantic gaze is given considerable treatment by Urry (1990) and is clearly both personal and collective, and highly pertinent to our analysis here. Romanticism is usually discussed in the past tense as a movement associated with the 19th century. We would argue, however, that as an aesthetic, *transcultural* practice and a discursive position, the Romantic view is alive and well as a present tense tourist way of being. Phenomena were anticipated through the Romantic gaze, perceived through the Romantic gaze, discussed in Romantic terms and were reinscribed, Romantically, into the public discourses on Scotland. As Arranz (2004) demonstrates, the marketing of Scotland as a destination relies heavily on Romantic images. Romanticism then gives tourists a common way of perceiving Scotland and of communicating Scotland, from their different cultural positions. Photographs evidence this.

Barthes (1977) famously examines photography as a semiotic connotation, supplementing signs by associations from culture, and an institutional activity integrating and reassuring as a practice. The analysis of photographs and their content has led to the development of a literature in tourism and, more widely, in cultural studies, regarding the interpretation and circulation of photographic imagery. It is not our aim to engage in semiotic analysis of photographs here. Rather we are interested in the practice of photography as an activity in the present, often framed Romantically and with transcultural communicative potential, that is oriented towards memory and towards narrative capital, as discussed earlier.

Photographs, produced in a particular place in response to the claim of phenomena and as a result of the democratisation of recording technologies, and conditioned as a social and cultural practice, are material objects. They are developed, usually – even with digital photography – after the moment of photographing. In other words there is a time lapse between the moment when the immediate scene claimed the photographer and the resulting photograph, which, as a text in its own right, then works semiotically, emotionally and materially to

Always a photo-opportunity: A highland cow

exert further claims on those who view it. In this respect the photographic images act like written narratives, ruptured from the immediate context and rendered mobile, and consequently available to a volatile range of emotional responses and imaginative possibilities. As such, photographs are examples of souvenirs that move between and nestle into different contexts, out of tourist time.

Collecting Culture

'Souvenir' of course, coming from the French 'souvenir', means not only keepsake and memory, but is also the verb 'to remember'. Photographs, as souvenirs, form part of a collection of artefacts that are taken home and that become the stimuli for memories and for stories. They are stored away in albums that are unique collections. The pause for the photograph looks forward to the future activities of cutting and pasting, sticking and labelling, which go alongside the development of an album and a photograph collection. This is an activity that Benjamin sees as characterised by the passion for collecting:

> There is no living library that does not harbour a number of book like creations from fringe areas. They need not be stick-in albums or family albums, autograph books or portfolios containing pamphlets

[...] For a collector's attitude toward his possessions stems from an owner's feelings of responsibility toward his property. (Benjamin, 1973: 68)

The other kinds of materials Benjamin mentions, pamphlets, leaflets etc., also get taken home, and brought out again at times triggered by other people's need or interest, again responding to and creating other journeys in the imagination that may be fed by images and mobile textual cultures. The emphasis on property and on the 'living' nature of the library is significant for our purposes here. Whatever was being collected, whisky, leaflets, information, photographs, the activity was concerned with making these belong, making them stick: 'the phenomenon of collecting loses its meaning as it loses its personal owner' says Benjamin (1973: 68).

The collecting we witnessed, and indulged in ourselves as part of our 'data collection', was involved in the making of meaning in the present for future purposes. It was another aspect to the tourist grammars we noted in Part 3. It evidenced the journey in material ways, it hinted at creativity, it allowed for reflexivity. It was emotional work, responding with the whole body to the call of scene and its future aesthetic and narrative potential. It was a social convention. But each click of the shutter promised differentiation in the albums and the stories back home:

Thus it is, in the highest sense, the attitude of an heir, and the distinguished trait of a collection will always be its transmissibility. (Benjamin, 1973: 68)

'Collecting' tourists we see as present to their environment, looking out carefully for artefacts and photo opportunities, yes, for the moment as part of a way of doing tourism but also for the future and for the stories these tell and the artefacts that they may become, once back home. By way of example, collections of holiday photographs of Skye were to be transmitted publicly again by our tourists in the form of lecture shows at the local evening class, websites, CD-ROMs and private albums.

What we see occurring here with images is the collector of photographs – and we are not claiming in any way that all are passionate photographers when on holiday – gathering materials that, once back home, serve to write culture and to reaffirm the primacy of written, textual narrative, over the oral narratives privileged when on holiday. These new mobile possessions, these tourist texts, are both personal and public. They retain an intimacy and the personal marks of perception and experience but they are often published, broadcast on websites, on sofas after dinner, in coffee bars in the lunch break, on desks. They are representations and possessions, and as such they become social facts.

> O bliss of the collector, bliss of the man of leisure! Of no one has less been expected and no one has had a greater sense of well-being than [...] a collector. [...] Ownership is the most intimate relationship one can have to objects. Not that they come alive in him; it is he who comes alive in them. (Benjamin, 1973: 69)

The set of object-relations acted out here is a complex one, attending both to the present and the environment, but also, importantly, to the future and to nontourist life. We may argue that, as an activity that is one of collection, photography enables the future layering of memories, of liminal, leisured times onto the routines of everyday life, back home. It enables the assurance of an intimacy with one's own memories, a reassurance that they are intact, evidence of their having-happened, all of which act as prompts for narrative, out of the chaos of memory.

Packing to Return

Perhaps as a general rule, the same kind of preparatory, anticipatory care is not taken over the packing for home as in the packing for the holiday. We know what we will need, and the majority of the items we have with us are actually ones that have their allotted place in the routines of the lives we lead back home. The activity of careful folding may be replaced by stuffing. New fragile purchases, however, are wrapped in worn clothes, sandwiched carefully into central compartments of bags or car boots, for fear of breakage or spillage.

Dirty washing is separated out, into side pockets, bags, compartments. The same order that applies to packing for the outbound journey does not necessarily apply to the return. Some things are ditched, never to return – old T-shirts that have given up the ghost, or been replaced by something with a snappier logo. Some foodstuffs and used pharmacy may become waste in the process of selection that accompanies the return packing.

More is known of the journey and its exigencies. The logic of the ordering of things for the return is based on the known more than on the unknown. 'Where are my house keys – I will need them to be handy? Careful with that bottle of whisky – that's for the neighbours – they've been looking after things for us.' The things that are now the 'most looked after objects' are the things that evidence the experience; the presents and the keepsakes, souvenirs, mementos – all aptly named for their ontological status as receptacles for memories, wrapped in smelly T-shirts.

It is not only objects that are packed. Different moods accompany the return from holiday. The flowing tourist bodies alive to new experiences are changed, muted, still responding but now to the pervasive sense of the return journey that manifests itself physically in the present. Emotions and sensations are packed as part of the preparation for a return to the familiar.

> The dorm is silent. Everyone is concentrating hard. We are all moving on today and it is hard work getting everything together again. There is little sense of anticipation, rather more an elegiac mood. 'Don't forget your coat'. The shouldering of packs, quiet smiles, a queue to sign out at reception before 10 a.m. and then farewells to those we have got to know. We shake hands, hug even, there are some gentle jokes to help the momentary awkwardness pass.

What we see in this extract is a representation of the emotional transition from a moment of communitas, as an intercultural phenomenon, to familiar cultural structures. The awkwardness points to these relations as strange relations, partings experienced for the first time, as oddities in Modernity, which rely on an alienated mode of social being. And the journey, as Turner and Turner (1978) point out, is not linear but elliptical. There is not one road but two. Within all the small arrivals at points or markers along the way – and there are many of these recounted in the different diaries of tours to Scotland and weblogs made by tourists – there is a sense of a wider journey unfolding.

This makes the *return* the 'real' destination. The aim is not so much to reach the holiday destination as to return home safely, with bags intact and stories at the ready. The non-places (Augé, 1995) of ferry terminals, hire car offices and airport lounges that characterise travel between Western nations and Scotland give way to familiar places, to home and to the routines and threads of nontourist life. These so-called non-places are nonetheless important and peopled places in and of themselves, finding their way into our return narratives of the journey home, providing opportunities for surprising encounters with friends who happen upon us, enabling a final set of metonymical souvenir purchases to be made. The predominance of the public, mythic scripts relating to Scotland in airport lounges, border service stations and ferry terminals all act as signposts for the reintegration of written, public narratives into everyday life.

And then we arrive back home.

Picking up the Threads

It is interesting to note that of the instrumental attributive adjectives in the grammars of travel, 12 refer to the dirt and disorder that come at the end of a journey.

> **5.** *attrib.* and *Comb.*, [...]instrumental, as *travel-broken*, *-disordered*, *-soiled*, *-spent*, *-stained*, *-tainted*, *-tattered*, *-tired*, *-toiled*, *-wearied*, *-weary*, *-worn* adjs.; (OED)

Travel, we are reminded, in English at least, grew from 'travail' – from labour, interestingly, not from the idea of rest. There is work in returning, just as there is work in resting. By and large we set off clean. We take a bath or a shower, change out of our work or everyday clothes into clean ones for travelling. We wash the clothes we wish to take with us, we fold them neatly for the purpose of packing. By and large we return home dirty. Travel-worn or travel-weary may indeed be phrases applicable to arrivals at holiday destinations, but the feelings diminish in the excitement of arriving, and the changing into clean clothes. They apply much more clearly to the return. The holiday is over, the return journey is complete and life resumes its normal course. Arrival home is typically of lesser, general narrative value than arrival in a new place. The journeys are the features. Far from having to invent the habitual anew, the habitual is merely dormant, and surrounded by familiar material effects.

Of course stories abound of failed re-entry into everyday life. A leak, a burglary, dead plants, bad news; all are possible disruptions to the smooth picking up of the threads of everyday life that could be sitting, out of time, waiting for time to begin again. All of the elements identified by Graburn (1978) as attendant upon the little deaths of our departures feature here – the instructions for watering the plants, for the care of pets, for the safe guarding of keys, the neighbours and family and friends into whose care some of the materiality of our lives are entrusted, however temporarily. Graburn also sees the return as problematic:

> The reentry is also ambivalent. We hate to end vacation, and to leave new-found if temporary excitement; on the other hand, many are relieved to return home safely and even anticipate the end of the tense, emotion-charred period of being away. We step back into our former roles often with a sense of culture shock. We inherit our past selves like an heir to the estate of a deceased person who has to pick up threads for we are not ourselves. We are a new person who has

gone through recreation and, if we do not feel renewed, the whole point of tourism has been missed. (Graburn, 1978: 23)

And there is routine business to deal with; in greater quantities, the pile of unopened post either by the letter box in a disorderly heap, or carefully accumulated on the dining room table, or by the telephone, by a kindly soul who has been in before us to check all is well. This helps the return, with some orderings of things – so as to minimise the sense of absence and of chaos.

Non-tourist Life

Drinking, eating, washing, sleeping, stack up as immediate routines back home. Phone calls may be made: 'we're back' – to our close friends and family. 'How was it?' 'It was nice'. Fragments of news and views and stories hastily exchanged, not in a long comfortable fireside chat, or over a leisurely meal, but as a perfunctory reassurance that things are well. Time and experience are condensed into fragments of speech and narrative, to be brought out again, perhaps, in more textualised forms later.

Everything is unpacked, things tidied away, homes found for new objects, purchased on the trip. Over the next couple of days photographs are put in for developing, or plugged into the computer. People are greeted again. Connection is re-established with the routines of life. We buy fresh food. We discover when it was our postcards arrived home. Gradually we may begin to make scrapbooks and albums of our trip. We find our tourists beginning to write culture. Websites abound in cyberspace with travelogues, stories and photographs of trips to Scotland.

In this respect we might understand our returning tourists, engaged in representational, writing cultural activities, as acting as cultural translators. They are rendering experience textually and using a wide range of material objects, images, souvenirs and, of course, language, to move meaning from past to present, from oral to written narrative, from place into representational space:

> Recording cultural stories in writing [...] fixes the storied events in their particularity, providing them with a new and unchanging permanence while inscribing them in a steadily accreting sequence of similarly unique occurrences. (Abram, 1997: 195)

Tourism leaves traces in the host culture and these are well documented in the tourism literature, but it also takes narratives back into the home culture for further acts of exchange. They create imaginative trails that

Saying goodbye to new friends outside a hostel

are evocative of feelings and play on the senses – the touch of the pottery souvenir, the play of colours from a new photograph, the recall of a foreign language phrase on the tongue. The metaphors and adjectives that are used to describe something of the synaesthesia of vivid perception and experience is often literary, or poetic, where, as Mcfague (1975) argues, metaphors become the means by which thoughts are able to move. They are the medium for the imagination.

The holiday has been a change, a break, a rupture to routine. For the kinds of tourists we have been encountering we can see, along with Graburn (1978), that this suspension of written narrative and textual practices in favour of an oral hierarchy represents a inversion of normal life. In this sense our tourists were, to some extent, following the kinds of inversions that are already well documented by anthropologists – that those with less economic capital tend to seek out material luxury and those with more seek out material simplicity.

This kind of change, or pause, makes space and time for other ways of doing life, ways we understand as doing tourism. The effects of this break may linger on afterwards in the stories that we tell, long after our clothes are ironed and folded away again. Much will be forgotten. But some stories will be favoured, brought out again and repeated time after time. They may be tales of disaster – that lost teddy bear again – or they may be humorous.

The end of the holiday means a packing away of collected experiences and experience. This involves memory and objects and narrative, interweaving. The time for displaying these again will depend upon circumstance and an order, and a form we may bring again in the future. The retelling of stories, the creation of logbooks, web pages, slide shows or just the taking out and turning over of stories from our memories is reflexive work, such as we described in Chapter 3. It occurs in space and time, as an embodied moment with glue and lamplight, with websites and the internet, over food and with friends and family.

Our favourite phenomena – stories, souvenirs, photographs – are left to rest, to recover from their travels and then brought out again, displayed both purposefully and spontaneously, but in new forms, as representations, and new orderings than those of their 'real' time experience. In this way we see the creative effects of a trip away brought back into the material contexts of our nontourist lives and the labyrinths of memory. In this we also see the limits to our structure in this book.

As we noted at the outset, the use of packing, unpacking and repacking is a device for ordering material. It works with the temporal direction of a holiday, from beginning to end, but what we see here is that actually these practical instances of packing and unpacking are not neatly constituted, but they overlie each other, nestling together. Some-one's story and practical experience, someone's maps, guidebooks and pamphlets, as we noted in Part 2, becomes someone else's packing. Whilst packing, unpacking and repacking might be conceptually separate, their temporal modes of past, present and future are in fact mutually constitutive.

The journey is indeed elliptical, and it shades into the routines of everyday life, but that life, while not radically reconstituted, does take on the possibility of different tones and hues as the experience is reconstituted in new forms.

It is here that we feel the weight of the anthropological tradition that sees tourism as a sacred journey (Turner & Turner, 1978) and as a liminal, ritual event. These are powerful and evocative ways of understanding tourism and they have a grand narrative function. We would broadly ascribe to their premise but would also wish to make a distinction between such strategic metanarratives of tourism and the more mundane tactics of tourists and their narratives. Tourists do not, on the whole, talk of their holidays as holy-days, after travelling to Scotland. Their discourse, if anything, is more Romantic than overtly religious, as we mentioned earlier. They may experience feelings of anxiety and culture shock, but these activities do not only have to be ordered into metanarratives of transition, of liminality, of symbolic death, even if we are able to discern such traits. Rather than wishing to ascribe wholesale to theories of the liminal, sacred, even transcendental aspects of tourism, we are attempting to work here with a more incarnational, material approach to the view of tourism as sacred journey.

The phenomena discussed in this chapter and throughout the book, their ways of claiming us as tourists and of being reciprocally claimed by us, do not speak so much of encounters with the transcendental gods, as of a participatory set of actions in a peopled, material world. These actions and claims may be dramatic, full of narrative potential, or small-scale ephemera, but they do not banish the liminal, liminoid or communitas to transcendental spaces. Rather these actions in the world, these reciprocal perceptions and acts of intercultural encounter render life tangible, touchable and sensible in fresh ways that punctuate and are part of everyday life.

Just as we noted that everyday life reinvents its routines in tourist spaces and in tourist time, so we discover the tourist space and time is reroutinised and embedded into the spatiotemporal rhythms of the ordinary, of non-tourist life. We are not dealing here with an either–or: clean:dirty, past:present, tourism:nontourism, sacred:secular binaries, but with a dialectic of practices and intensities between different modes of being. In other words, the sacred journey is rendered ordinary. Tourism, we are arguing, changes our relations to time and space and place and our awareness of phenomena in such a way as to make this possible.

Chapter 11
Conclusions

Tourism matters. Why it matters and how, in particular instances of tourism as a mode of being in the world, has been the focus of this book. Key to our endeavour has been the concept of exchange, whose dimensions and machinations we have studied with particular relation to the vicissitudes of the intercultural life of tourism. We have viewed exchange not only as an important social practice of tourists, but also as a wider metaphor for understanding tourism as an everyday social activity. In this final chapter we draw together some tentative conclusions from our attempts to present a theoretically driven, empirically grounded study of the intercultural life of exchange before, during and after travel.

The importance of understanding tourism in this way, we shall argue, lies in the manner in which tourism has the potential to teach us some radical lessons about the possibility of living a different, everyday life when we are not tourists. For within tourism there are experiences through which a capacity to tell different stories about ourselves and others may be called into play. The manner in which we pursue lives dialectically with others in the socioeconomic context of (late) modern capitalism can be the focus of these stories and can provide a jolt to the contemporary visibilities and invisibilities of our everyday ways of being.

Playing with the Material and Metaphorical Life of Exchange

We have chosen to frame our presentation of the intercultural life of exchange through a practical and metaphorical examination of packing, unpacking and repacking baggage. Baggage and packing practices mark us out as tourists. They are words we use of our possessions only when we are on the road. We are aware that there is a degree of potential postmodern play involved in both the material and the metaphorical use of the language of bags and packing here. As in the tourism literature so too in the literatures of criticism and hermeneutics, we frequently encounter the dichotomous suggestion that readings of texts and culture(s) must be either postmodernist and playful, or historical and material. We would like to suggest, however, that there is more to our material metaphors than simply fun with words. Our investigations point to the historical and the material, as much as the playful and the ironic.

The dominance of varying appropriations of deconstructive readings of text and culture, from the mischievous to the political, does not mean that historical and material concerns have not continued to influence hermeneutic practice. Valentine Cunningham (1994) launches a passionate attack on the dominance of certain renderings of deconstruction and the postmodern, and seeks to redress the balance of literary work. He maintains that history and the material contexts of the world are a fundamental element in the work of texts and of reading:

> My argument is that we not only need not choose [between empiricism and deconstruction], but that rigorous analysis will not allow us to choose: *il n'y a aucun sens* of doing otherwise. The true Saussurean, the true Derridian, the linguistic case is that meaning arises at that duplicitous, slippery place where apparent opposites apparently conjoin, so that both of the connecting, opposed sides of that border must inevitably get taken and be read and interpreted conjointly. This is the logic of the betweenness of writing, of works of art, that Martin Buber has classically endorsed: 'Art is neither the impression of natural objectivity nor the expression of spiritual subjectivity, but it is the work and witness of the relation between the *substantia humana* and the *substantia rerum*, it is the real of "between" which has become a form.' (Cunningham, 1994: 60)

When we come to use the playful language of packing and baggage we are attempting to reflect on how meaning itself is thought and practised through the material forms of 'between'. As we outlined in Chapter 2, drawing upon the work of Celia Lury, tourism research, through anthropological research more generally, has only recently tackled this issue of the in-between, primarily through the idea of the liminal as a ritual practice. In examining the emergent material life of tourists' 'dwelling-in-travel', we have attempted to take the elision of this conceptual distinction seriously by grounding its discussion in the everyday life of the material and cultural realities of the travel bag metaphor.

At one level, such empirical study of the dwelling–travelling cultures of the people and objects that constitute forms of tourism has assumed, as much as it has demonstrated, the flimsiness of the many dualisms that characterise tourism research, particularly those anchored within some of the grand narratives outlined in Chapter 2. Our research attests to the difficulties of confining understandings of tourism to the commodified exchanges of an abstracted market through which consumers are either passively assimilated by the interests of capital, or actively cynical towards them. In unpacking our own travel bags, and observing the unpackings of others, we have come to understand that tourism criss-crosses the borders of work and leisure, of gifting and commodity exchange, of dystopia and utopia, of assimilation and cynicism.

At another level, we found that the layering of the material and the metaphorical was articulated around a variety of divergent and locally embedded forms of exchange. By paying attention to the local contexts of dwelling and travelling, it is possible to see how the myriad of nonmarket exchanges, and their social relations, put the commercial firmly 'in place' and demonstrate that this sphere is not present at all times and in all spaces. There were times when tourists were inter-pellated by the calls and practices of commodified exchange, and those when they were not, as much consumers as producers of everyday life and exchange. Let us consider this in more depth.

What facilitates or hinders intercultural communication, and the role of exchange within this, is a question that has guided this book. Exchange was crucial to the fate of intercultural communication on holiday in a number of contextually specific and politically interesting ways. What we found on the Isle of Skye was an array of alternative manifestations, and intensifications, of practices of exchange: narrative, economic, bodily, material, intercultural, of knowledge. Box 11 illustrates the different manifestations of exchange mentioned in previous chapters.

Exchange was, then, a key practice of everyday tourist life. It took different forms, as illustrated below, and was therefore constitutive of

Box 11 Exchange as ...

Narrative	Material	Intercultural	Economic
Temporal	Object-relations	Help	Monetary
Material	Hermeneutics	Proximity	Narrative
Social	Props for story	Food	Help
Didactic	Telling	Tips	Information
Capital	Bodily cleansing	Information	Food
Spatial	Food	Listening/telling	Consumption
Bodily	Consumption	Translating	Rehabitualised action
Recording	Rest	Languages	
Revising	Reflexivity		
Practising			

different modes of dwelling-in-travel. Sometimes it materialised in market-based exchange in tourist bubbles; sometimes it took the form of swapping food in youth hostels where the currency was a smile; sometimes it involved the telling of stories and the swapping of tips and information. Exchange was also an imaginative and an emotional act. For the Germans we encountered in Chapter 8, for instance, it involved working within the realms of the imaginative architecture of the Romantic myth of the north, and exchanging resultant expectations and inspirations, for the vagaries of material reality.

And for us, unpacking our travel bags on arrival at our temporary destination, as Part 3 all too painfully illustrates, was something of an emotional jolt for the uninitiated: feelings of being out-of-place, and away from home, encoded in the material objects of a hostel and enacted in the uncertainty of learning emergent codes of behaviour. These different modes of exchange, and associated emotions, meant that in some places and at some times, the social relations necessary for intercultural communication were occluded, whilst in others they flourished, as Chapter 9 illustrated. These contextually specific modes of exchange had politically interesting consequences indexing different relations to capital.

What we have described, from our data and our theoretical frameworks, is exchange in action, a set of tourist practices in which power and competition is not *always* the dominant rule for social exchange. On the one hand, we did participate in the interpellation and embodiment of the strategies of capital, to use de Certeau's term. This was perhaps best illustrated in the cases of the Flora Bay and Peaty Distillery through the docile bodies, us included, of the hotel guests and the guided tour. In these cases the frustration of interaction and dialogue was 'worked at' by all involved, and crucially monitored and regulated by the guide. It might be said that both these contexts are part of an archetypal tourist-scape where commodity relations dominated and impacted negatively upon spontaneous and unscripted social interaction.

But on the other hand, to repeat, this was not always the case. For what was revealed through a giving of ourselves to a process of observation and participation, is that rules of exchange are not simply dominated or exclusively circumscribed by tourist-scapes, tourist traps and capitalist ventures. We 'did' tourism, 'in and out' of capitalist relations of exchange. Our experiences, as detailed in several sections throughout this book, told of emergent rules and structures of exchange based on different values, and of resistance to the commodity relations that are very often assumed to inhabit the totality of tourist exchanges.

The multiple forms of exchange in which we participated demonstrate that capital is far from an omnipotent force in tourist life. They instantiate de Certeau's notion of the tactic, that is, everyday resistance to the strategies of capital, and puncture a hole in any presumed hegemony of capital. Such 'alternative', in the sense of nonmarket, practices of exchange, provide the basis for alternative stories to be told which do not automatically cast tourism as the inevitable outcome of relations of production and consumption.

Instead, the variety of different exchange practices we observed or participated in gave material form to the in-betweenness of tourist life. And the contiguities that characterised this in-betweenness suggest that tourism should be thought of in terms of both–and, rather than either–or. In moving towards the close of this book, we choose to dwell on the alternative stories about tourism that our study leads us to. Such alternatives emanate from an empirical study that has tried to work critically with as well as against the grain of the universal stories of previous theorising. These alternatives are the fruit of our own material and metaphorical play with the material and metaphor of the travel bag.

The Abundance of Tourism: An Alternative Story

What is it that makes our story of tourism an alternative one? One reason, mentioned earlier, is that it demonstrates that tourism is neither good nor evil, alienating nor emancipatory, purely commodity exchange or gift, material or metaphor: it has the potential to be each of these things. Tourists work through these oppositions: they become a productive and a consumptive form of life, part of the complex miracle of the mundane. We would like to suggest, however, that our alternative has further dimensions. More specifically, it tells of the ways in which the intercultural life of exchange can be one of abundance, that exposes in fundamental ways the dialogic principle of relating to others, and undermines the myth of scarcity that drives contemporary economic life. Tourism matters because it can make visible the dialogic principles that open our eyes to the very abundance of human life. It is this potential lesson of social bondedness, of sharing process and phenomena, which we need to take back to our nontourist life.

These conclusions about the reinvigoration of forms of social bondedness in times of holiday, and of the abundant phenomena that claim us on the road, can be read back through the interpretive insights of travel liturgy, often associated with the wider notion of tourism as sacred journey (c.f. Bauman, 1996; Graburn, 1978; Turner & Turner, 1978).

Perhaps surprisingly, casting our net outwith the confines of the tourism literature on sacred journey in order to develop these issues, took us to interesting and illuminating theological work on the relationship between travel/journeying and social relations/community. In this respect, the works of Rowan Williams (2000) and Walter Brueggemann (1999) were suggestive of fruitful ways to open out our understandings.

To begin with the context of cultural history, Williams (2000), through his reading of Bossy's work (1984) on cultural change in medieval church and society, provides a first point of resonance and development. His commentary suggests how in (religious) holidays of times past, alternative social relations were made possible as the frames of references of competing communities were worked through in renewed ways and with changing relations to the body, through the suspension of everyday rivalries. In his text, Williams, influenced by Foucault, is concerned with the shifts that take place in how we understand and practice social bondedness. In the Middle Ages, the summer holiday of Corpus Christi was a time when rivalries were postponed, mystery plays performed and there were public processions of the Sacrament of Christ's body. In such contexts Williams (2000: 68) describes how:

> social meaning works, by a kind of 'nesting' of frames of reference within each other: the social body, the Church as Body of Christ, the sacramental presence of Christ's body in the Eucharist; a subtle crossing and re-crossing of the boundaries between fields of discourse.

This is an instance where, to repeat Cunningham, 'the "real" of between has become form'. In-betweenness is a position; not a significatory drain through which social life falls away from us. It is a practice of everyday life and bodily practice too. Here we can augment the rather constrained view of 'between' in the tourism literature, which we mentioned earlier. To reiterate, the in-between is usually cast in terms of the ritual practices of liminality, part of the sacred and transcendental aspects of life. What we are suggesting, however, is that negotiation of the in-between involves material, imaginative and creative forms (be they matches, green peppers or diaries for our recollections). The 'sacred' is therefore an ordinary and everyday experience, perhaps not so much a transcendental as an incarnational aspect to tourism. The characteristic ambiguity of the liminal is not just part of a 'marked-out' ritual process; it is also an 'unremarkable' facet of the intercultural life of exchange.

The nesting of frames of reference of different tribes and their social relations in times of festive nonrivalry provided the potential for new forms of social bondedness like those around the feast of Corpus Christi. There are clear parallels to be drawn here between the sharing of discourse in times of peace and noncompetition, and the forms of communitas that marked the creative exchanges of places like the youth hostel. Our own study obviously occurs, takes form even, within a markedly secular age and yet it evokes earlier practices of the Sabbath, of holy days, holidays and of modes of life that were sustained prior to capitalism, globalisation and neoliberalism. We do not mean to adopt a Romantic view of times gone by, and of the history of the church in particular. Both are replete with their own examples of abuses of power and exploitations of people. What we are pointing to here is a parallel between the way in which changes to existent and competitive social relations became possible, both in times past through the institutionalisation of holidays (Sabbath days) by the church, and in more contemporary times through the particular practices of tourism suggested by this study.

What we find in our alternative stories is that the rules of rivalry and competition and anti-neighbourly practices, that fundamental basis of Modernity's story of the tourist and of capitalism, did not apply in our alternative tourist economies. It is during times of holiday that our tourists exchange the stuff of everyday life for a lighter set of rules and objects. It is during these times that more space and time is given to attending to their bodies, to sharing their goods, their knowledge, to giving help and advice, to telling stories. The process of packing is specific, has certain cultural quirks. But it is also a ritualised way of suspending the routines of everyday life and trying out other ones, taking things out of the bags of social life, in a new place, and ordering them in different ways, however temporarily, and without forgetting the rules of the nontourist game. It is during the times of holiday – Sabbath times, Seventh days of Seventh months – that these 'nesting' frames of reference are most clearly demonstrated in the Middle Ages, around a concern for the Body – social, political, Eucharistic – made manifest in a different, celebratory practice of its habits.

In the Middle Ages, this involved the nesting of the competing values and lives of warring factions; in contemporary tourism, it involves the nesting together of culturally and linguistically diverse travellers. The rules attendant upon capital were suspended, as a Sabbath principle, enabling, often through the action of normative communitas as in the youth hostels, what Turner refers to as 'flow':

> [...] What I call communitas has something of a 'flow' quality [...] it does not need rules to trigger it off. [...] Again, 'flow' is experienced

within an individual, whereas communitas at its inception is evidently between or among individuals – it is what we believe we share and its outputs emerge from dialogue, using both words and non-verbal means of communication, such as understanding smiles, jerks of the head, and so on, between us. (Turner, 1982: 58)

These times of 'flow' and of communitas, of exchange, as Turner describes them, are *not*, we would argue strongly, a constant, sustained feature of practising tourism, but they may and do occur through its practice. This is a subtle but important point. In this way, and through the work dedicated to preparation and anticipation, these times tell us something of nontourist life. Such times of communitas can give rise to a reflexivity towards alternatives. Our tourists change the practices of the capitalist game and in the process of doing so, they unsettle the foundations from which capitalism's rules are made possible. Such a characteristic of the moments of communitas outlined in previous sections occurs precisely by slowing down its pace, by resting or stretching the body, by rupturing routine, pausing, inserting a *halt*.

Of key importance, as already mentioned, is the body, and the manner in which the 'nesting of frames of reference' of which Williams speaks, are conducted in and through flesh, and our attention to it. It is perhaps no surprise then that Chapter 7 should focus on the body out-of-habit – here, this stop, or this pause was a profoundly embodied, and emotional experience. In that chapter we described how our bodies fell out of routine, and how, through object-relations, we were required to reinvent our habits by sharing time and space with others who were unknown to us. As we described, it was precisely through the intercultural encounters in the youth hostel, through the renegotiation of our bodily routines, and the subsequent mixing of our frames of reference, that communitas was to occur, that sharing became possible. This 'halt' was a material form of the between and has a number of important qualities that can be articulated in further depth through the dialogical principle, or the principle of 'I–Thou'.

In *The Covenanted Self*, Walter Brueggemann (1999) explores the notion of covenant as a mode of shared life in a selection of Old Testament texts, arguing that covenant represents a radical alternative to what he calls 'consumer autonomy'. At the heart of shared life is the other, and Brueggemann defines this other not just as a counter object to the self, but as a verb. He writes that other is:

the risky, demanding, dynamic process of relating to one who is not us, one to whom we are accountable, who commands us, and from whom we receive our very life. (Brueggemann, 1999: 1)

Such a view is part of the principle of 'I-Thou' (also called the principle of alterity, or the dialogical principle), most often associated with the work of Martin Buber, mentioned earlier in the book, according to which there is no such thing as an independent and autonomous human agent. Rather our notion of self is coextensive with the notion of others, and the process of living therefore becomes one of:

coming to terms with this other who will practice mutuality with us, but who at the same time stands in an incommensurate relation to it. It is the tension of mutuality and incommensurability that is the driving force of life. (Brueggemann, 1999: 2)

Our time in youth hostels, and in other spaces of creative exchange and spontaneous interaction, where we reinvented our everyday routines, are precise reminders of the fragility of our selves and our bodies, and of the power of this dialogical principle. Faced with new people, new events, unwritten rules and emergent identifications, these various others reminded us that we are 'addressed, unsettled, unfinished, underway, not fully whom we intend or pretend to be' (Brueggemann, 1999: 2). They claimed us, just as much as we claimed them. Their claims on us, and the feelings of nervous excitement and trepidation they engendered, called for new modes of sharing life, or new 'covenanted' arrangements.

When we are tourists then, just as when we are researchers, certain phenomena make a particular claim upon us. They work upon us emotionally, intellectually and materially, drawing us into relations of exchange that are often involuntary and unpredictable. We should be clear, however, that these phenomena, these others, are not only physical human beings; these others are also events, occurrences, history, stories, material artefacts *inter alia*. Furthermore, it should be noted that we are not claimed as virgin territory. Like the phenomena we encounter, we are made up of prior sets of traditions, histories, disciplinary practices. Such phenomena, in their choosing of us, can call us out of established routines and everyday modes of living into different, even novel, ways of being, in alternative conversations.

In the give and the take that characterised the emergence of such conversations and covenants, we practised a dialectical move of self-assertion and self-abandonment; simultaneously making arrangements to meet our needs and to call on others to assist in this, whilst giving up on certain practices and ideas as we were called on by others. Such a covenanted self, based on what Brueggemann calls a 'dialectic of reconciliation', suggests to us that tourism might matter, as it can consist of 'learning the skills and sensitivities that include both the courage to assert self and the grace to abandon self to another' (Brueggemann, 1999: 8). It is the other that demands a halt, demands a rupture in our routine. It is in the principle of 'I–Thou' that alternatives can be found, alternatives to our taken-for-granted ways of doing things based on an awareness that the either–or choices we make could be reversed and life could be practised differently with others.

Tourism could therefore be both a rhythmic intensification of the same, as much as it could be a time that is 'out of joint' in which alternative social relations become possible. Tourism could be a site for the tactics of otherness. We would suggest that tourists live their everyday lives 'in-between' these two possibilities, shifting between one and the other, and finding material and emotional comforts and discomforts in these moves. It is in this realisation of the in-between, the continuing transgression of cultural categories and its manifestation in the material, historical and emotional life of tourism, that we detect a clue as to why it is that tourism matters. In short, it can bring to light the very provisionality of our world, its articulation in our everyday social relations and a vision for a more abundant everyday locked into the principle of alterity.

Some travel is explicitly undertaken for reasons of self-discovery and pilgrimage, perhaps to escape self and its relations to capital. Some travel is undertaken as a continuation of everyday forms of consumption, commodity forms exchanged for disposable income. In our travel, the practices of liminality that characterise such continuities and disconti-nuities are not rituals disengaged from our everyday selves. Instead they are dialectics that draw on yet diverge from our collective 'every-daynesses' and engender the possibility of destabilising our relations to self. Such dialectics are forms of lived in-betweenness, and they can contain elements of self-discovery that jolt us out of the tacit commodi-fied forms and taken-for-granted routines that often frame our relation to others. Tourism, as much as it has the capacity to promote sameness and continuity, also has the capacity to spell difference and change.

We are not saying that tourism gives us a glimpse into alternative social relations that require a renouncement of self. This would be undialectical, and likely impossible too. Instead, the alternative lies precisely in the dialectic of sharing life, this dialectic of reconciliation, wherein lies, to paraphrase Brueggemann (1999: 19), a set of social practices that overcome 'fearful conformity and troubled autonomy'. The importance of the kinds of aspects of intercultural tourist life we have identified lies in the dialectic of self and other, in the tension between the two, a tension that provides the necessary destabilising force to remind us of the choices we make and of the force of history. This tension concerns moments in life when we become, following Kristeva (1991), 'strangers to ourselves'. These were moments of rest, moments of sleep, moments of food, moments of cleansing and attending to the needs of our bodies. We need to dwell in the tension that lies between, and is materialised in the dialectic of identity and self-assertion and the giving up of stuff. It is an everyday working ambivalence of Modernity, of the sort described by Wang (2000). Such a dialectic suggests that in tourism, there is an important space for an examination of social bondedness and social meaning. It is the forms of spontaneous, ideological and normative communitas mentioned earlier, that 'nest' our frames of reference and our practices in tourist life, that make tourism matter.

There is a second point of resonance between the alternative stories of tourism we are telling, and the covenanted self discussed by Brueggemann. Tourism is not just a challenge to the self through others; it is also a challenge to one of the basic tenets of economic life. Brueggemann argues that the myths of scarcity, upon which capitalism is founded and which can lead to deeply unethical practices of life, may be challenged by the telling of alternative stories and the development of alternative practices of life and exchange. In our alternative stories and instances of communitas we find a practice of *abundance* – there was enough time, space, knowledge – for doing tourism together. We borrowed each others' food, shampoo, time, pens, hopes. There was plenty for all. These practices of exchange were fruitful and neighbourly, small and do-able, within the regimes of tourist life, even if they are not apparently sustainable in nontourist life. Brueggemann describes this kind of abundance, the notion that there is more than enough in life, as one of the key messages of the Old Testament. It is a recurring message of generosity that is perhaps best exemplified in the story of Exodus (the Biblical word for departure). As an enduring and important story of a sacred journey in Western culture, Exodus provides a way of opening out a core understanding of tourism in anthropology.

The story of Exodus was not just a geographical event, according to Brueggemann. It was also an 'economic act, an imaginative act' (Brueggemann, 1999: 113) aimed at breaking away from the myth of scarcity that determined economic life under the Pharaoh. The story of Exodus aimed at undermining a myth at the core of which lay fear and anxiety, rather than rigorous economic analysis. It aimed to rupture an ideology that turned humans into 'agents of acquisitiveness in the face of all others who also pursue acquisitiveness' (p. 112) and to live outwith this ideological claim. According to Bruggemann's interpretation, the story of Exodus represents the central work of Israel as the establishment of a 'counter-principle of life and faith that refuses, from the bottom up, the assumptions, claims and persuasion of Pharaoh' (p. 113). This counter principle sought a particular kind of covenant, or mode of sharing, based on generosity and abundance, and not the erroneous and callous myth of scarcity.

Our alternative story of tourism reminds us of the possibilities that social and economic life could be organised around a notion of abundance rather than scarcity. We should be clear to say however that our experience of tourism was not solely one of new and emergent forms of sharing that provided a basis for intercultural communication. This was not true and, of course, is undialectical. We also found practices of scarcity where space was highly managed, scripted and where the opportunity of intercultural communication and for developing alternative practices was thwarted through a rigorous imposition of rules. And we should also be clear to say that in other contexts, such as rich Western tourism to areas of poverty, the 'material abundance' that provides the possibilities for travel in the first place rarely leads to an abundance of social exchange between tourists and locals. Abundance, in this case, can lead to limitation and ignorance. This is another context and the focus of other studies.

In our case, we suggest that our study has identified the dialectical possibilities for experiences of the intercultural life of tourism to be both limited and abundant, and as a result, to invert the relationship between abundance and limitation mentioned above. The travel bag – itself a material and symbolic form of 'limitation' – proved to contain more than enough to sustain our lives on the road. It compelled us to decide in advance, and place limits around our anticipated needs. But rather than creating further need based on lack, these limitations on our baggage were transformed into a life of plenty. Paradoxically, then, limit can create abundance. Limit does not have to be feared; it can also be perfectly fulfilling, if we believe it to be so, and practise limit as such. We might thereby teach ourselves of the possibility of practising and living limit as a form of abundance.

Tourism Matters Because of its Lessons for Everyday Life

Not being a tourist and being a tourist are not neat divisions of modes of being. Rather they cohabit in differing degrees of intensity. The liminal, communitas dimensions of our lives are always just within reach, spontaneously, ideologically and normatively. Sometimes these dimensions can be intensely ordinary, sometimes extra special. Tourism can be one of those ways of being where such dimensions nest together, accorded the time and space for a reinvention of the habitual and a reroutinising of the body. This change is indeed as good as a rest. Travel can broaden the range of our ways of practising life.

Tourism matters because it has the capacity to suspend everyday life and to suggest its potential. Tourism as a mundane practice shows us that life does not have to be exotic to make big changes in how we share it, and in how we think of our selves. Our lives just need to be jolted a little, and to be anything but luxurious, to give us a glimpse of our vulnerabilities to others, and the abundances of a life given to us by them. These glimpses, made visible through the dialectical tensions of moments of communitas, need to be remembered and to take form on return home. They might allow us to practise social bondedness differently, and to remind ourselves, in times of an increasingly but not all-pervasive capital, of the social miracles that attend struggles to come to terms with our need to assert but also abandon ourselves in the sharing of life. Such is the lesson of the intercultural life of exchange. Such is the reason that tourism matters.

Bibliography

Abram, D. (1997) *The Spell of the Sensuous: Perception and Language in a More-than-human World*. New York: Vintage.

Agar, M. (1994) *Language Shock: Understanding the Culture of Conversation*. New York: William Morrow.

Agar, M. (2000) *The Professional Stranger: An Informal Introduction to Ethnography*. London: Academic Press.

Alneng, V. (2002) The modern does not cater for natives: Travel ethnography and the conventions of form. *Tourist Studies* 2 (2), 119–142.

Anderson, B. (1991) *Imagined Communities: Reflections on the Origin and Spread of Nationalism*. London and New York: Verso.

Appadurai, A. (1986) *The Social Life of Things: Commodities in a Cultural Perspective*. Cambridge: Cambridge University Press.

Archer, M. (2000) *Being Human: The Problem of Agency*. Cambridge: Cambridge University Press.

Arranz, J.I.P. (2004) Two markets, two Scotlands? Gender and race in STB's 'othered' Scottishness. *Journal of Tourism and Cultural Change* 2 (1), 1–23.

Augé, M. (1995) *Non Places: Introduction to an Anthropology of Supermodernity*. London and New York: Verso.

Austin, J.L. (1975) *How to do Things with Words*. Cambridge, MA: Harvard University Press.

Barnett, R. (2000) *Realizing the University in an Age of Supercomplexity*. Buckingham: Open University Press.

Barnett, R. (2003) *Beyond all Reason: Living with Ideology in the University*. Buckingham: Open University Press.

Barthes, R. (1970) *S/Z*. Paris: Editions du Seuils.

Barthes, R. (1975) *S/Z*. London: Cape.

Barthes, R. (1977) *Image-Music-Text*. Glasgow: Fontana.

Bartlett, T. (2001) Use the road: The appropriacy of appropriation. *Language and Intercultural Communication* 1 (1), 21–40.

Baudrillard, J. (1993) *Symbolic Exchange and Death* (I.H. Grant, trans.). London: Sage.

Bauman, Z. (1996) From pilgrim to tourist – or a short history of identity. In S. Hall and P. du Gay (eds) *Questions of Cultural Identity* (pp. 18–36). London: Sage.

Bauman, Z. (1998) *Globalization: The Human Consequences*. Cambridge: Polity Press.

Bauman, Z. (2000) *Liquid Modernity*. Cambridge: Polity Press.

Bauman, Z. (2002) *Society under Siege*. Cambridge: Polity Press.

Bausinger, H. (1991) *Reisekultur. Von der Pilgerfahrt zum modernen Tourismus*. Munich: C.H. Beck.

170

Benjamin, W. (1973) *Illuminations*. London: Fontana.

Benjamin, W. (1982) *Das Passagen-Werk*. Frankfurt am Main: Suhrkamp.

Benjamin, W. (1999) *The Arcades Project*. Cambridge, MA: Belknat.

Bepler, J. (1994) The traveller-author and his role in seventeenth century German travel accounts. In Z. v. Martels (ed.) *Travel Fact and Travel Fiction: Studies on Fiction, Literary Tradition, Scholarly Discovery and Observation in Travel Writing* (pp. 183– 193). Leiden: E.J. Brill.

Bohannan, P. (1955) Some principles of exchange and investment among the Tiv. *American Anthropologist* 57, 60– 69.

Bohannan, P. (1959) The impact of money on an African subsistence economy. *The Journal of Economic History* 19 (4), 491– 503.

Boorstin, D. (1961) *The Image: A Guide to Pseudo-events in America*. New York: Harper Row.

Bossy, J. (1984) *Christianity in the West, 1400–1700*. Oxford: Oxford University Press.

Bourdieu, P. (1984) *Distinction: A Social Critique of the Judgement of Taste*. London: Routledge.

Bourdieu, P. (2000) *Pascalian Meditations*. Cambridge: Polity Press.

Brewis, J. and Jack, G. (2005) Pushing speed? The marketing of fast and convenience food. *Consumption, Markets & Culture* 8 (1), 49– 67.

Brougham, J.E. and Butler, R.W. (1975) *The Social and Cultural Impact of Tourism*. Department of Geography, University of Western Ontario.

Brueggemann, W. (1999) *The Covenanted Self: Explorations in Law and Covenant*. Minneapolis: Augsburg-Fortress.

Brueggemann, W. (2000) *Texts that Linger, Words that Explode*. Minneapolis, MN: Augsburg-Fortress.

Bryman, A. (1995) *Disney and his Worlds*. London: Routledge.

Buber, M. (1954) *Die Schriften über das dialogische Prinzip*. Verlag Lambert München: Schneider.

Buzard, J. (1993) *The Beaten Track: European Tourism, Literature and the Ways to Culture, 1800–1918*. Oxford: Clarendon.

Chaney, D. (1993) *Fictions of Collective Life: Public Drama in Late Modern Culture*. London: Routledge.

Clifford, J. (1988) *The Predicament of Culture: Twentieth Century Ethnography, Literature and Art*. Cambridge, MA: Harvard University Press.

Clifford, J. (1992) Travelling cultures. In L. Grossberg, C. Nelson and P. Treichler (eds) *Cultural Studies* (pp. 96– 116). New York: Routledge.

Clifford, J. (1997) *Routes; Travel and Translation in the Late Twentieth Century*. Cambridge, MA: Harvard University Press.

Clifford, J. and Marcus, G.E. (ed.) (1986) *Writing Culture: The Poetics and Politics of Ethnography*. Berkeley and Los Angeles, CA: University of California Press.

Conquergood, D. (1991) Rethinking ethnography: Towards a critical cultural politics. *Communication Monographs* 58, 179– 194.

Cooper, R. and Law, J. (1995) Organization: Distal and proximal views. *Research in the Sociology of Organizations* 13, 237– 274.

Crick, M. (1996) Representations of international tourism in the social sciences: Sun, sex, sights, savings, and servility. In Y. Apostolopoulos (ed.) *The Sociology of Tourism* (pp. 15– 50). London and New York: Routledge.

Cronin, M. (2000) *Across the Lines: Travel, Language and Translation.* Cork: Cork University Press.

Cunningham, V. (1994) *In the Reading Gaol: Postmodernity, Texts and History.* Oxford: Blackwell.

Dann, G. (1996) *The Language of Tourism: A Sociolinguistic Perspective.* Wallingford: CAB International.

Dant, T. (2000) Consumption caught in the cash nexus. *Sociology* 34 (4), 655–670.

Davies, A., Criper, C. and Howatt, A.P.R. (eds) (1984) *Interlanguage.* Edinburgh: Edinburgh University Press.

Davis, J. (1992) *Exchange.* Minneapolis: University of Minnesota Press.

de Certeau, M. (1984) *The Practice of Everyday Life.* Berkeley, CA: University of California Press.

Derrida, J. (1976) *Of Grammatology* (G.C. Spivak, trans.). Baltimore, MD: The Johns Hopkins University Press.

Desmond, J. (2003) *Consuming Behaviour.* Basingstoke: Palgrave.

Douglas, M. (1966) *Purity and Danger: An Analysis of the Concepts of Pollution and Taboo.* London and New York: Routledge.

Douglas, M. and Isherwood, B. (1978) *The World of Goods: Towards an Anthropology of Consumption.* London: Penguin.

du Gay, P. and Pryke, M. (2002) *Cultural Economy.* London: Sage.

Eagleton, T. (2000) *The Idea of Culture.* Oxford: Blackwell.

Edensor, T. (1998) *Tourists at the Taj: Performance and Meaning at a Symbolic Site.* London: Routledge.

Eriksen, T.H. (1995) *Small Places, Large Issues: An Introduction to Social and Cultural Anthropology.* London: Pluto Press.

Fabian, J. (1983) *Time and Other: How Anthropology Makes its Object.* New York: Columbia University Press.

Feifer, W. (1985) *Going Places.* London: Macmillan.

Ferguson, R. (2001) *George MacLeod: Founder of the Iona Community.* Glasgow: Wild Goose Publications.

Foucault, M. (1978) *The History of Sexuality. Volume One: An Introduction.* New York: Vintage Press.

Foucault, M. (1980) *Power/Knowledge.* New York: Pantheon Books.

Foucault, M. (1991) *Discipline and Punish: The Birth of the Prison.* London: Penguin.

Franklin, A. and Crang, M. (2001) The trouble with tourism and travel theory? *Tourist Studies* 1 (1), 5–22.

Fussell, P. (1980) *Abroad: British Literary Travelling Between the Wars.* Oxford: Oxford University Press.

Geertz, C. (1973) *The Interpretation of Cultures.* London: Fontana.

Glendenning, J. (1997) *The High Road: Romantic Tourism, Scotland, and Literature, 1720–1820.* London: Macmillan.

Graburn, N. (1978) Tourism: The sacred journey. In V. Smith (ed.) *Hosts and Guests: The Anthropology of Tourism* (pp. 17–31). Oxford: Blackwell.

Gregson, N. and Crewe, L. (1997) The bargain, the knowledge and the spectacle: Making sense of consumption in the space of the car boot sale. *Environment and Planning D: Society and Space* 15, 87–112.

Hall, S. and du Gay, P. (1996) *Questions of Cultural Identity.* London: Sage.

Hammersley, M. and Atkinson, P. (1983) *Ethnography: Principles in Practice.* London: Tavistock.

Ingold, T. (1993) The art of translation in a continuous world. In G. Pálsson (ed.) *Beyond Boundaries: Understanding, Translation and Anthropological Discourse* (pp. 210–230). Oxford: Berg.

Ingold, T. (ed.) (1994) *Encyclopedia of Anthropology: Humanity, Culture, Social Life.* London and New York: Routledge.

Ingold, T. (2000) *The Perception of the Environment: Essays in Livelihood, Dwelling and Skill.* London and New York: Routledge.

Jackson, P., Lowe, M., Miller, D. and Mort, F. (eds) (2000) *Commercial Cultures.* Oxford: Berg.

Jaworski, A., Ylänne-McEwen, V., Thurlow, C. and Lawson, S. (2003) Social roles and negotiation of status in host–tourist interaction: A view from British television holiday programmes. *Journal of Sociolinguistics* 7, 135–163.

Keesing, R.M. (1994) Radical cultural difference: Anthropology's myth? In M. Puetz (ed.) *Language Contact and Language Conflict* (pp. 3–24). Amsterdam and Philadelphia, PA: John Benjamins.

Kirschenblatt-Gimblett, B. (1998) *Destination Culture: Tourism, Museums and Heritage.* Berkeley, CA: University of California Press.

Kopytoff, I. (1986) The cultural biography of things: Commodization as process. In A. Appadurai (ed.) *The Social Life of Things: Commodities in a Cultural Perspective* (pp. 64–94). Cambridge: Cambridge University Press.

Koshar, R. (2000) *German Travel Cultures.* Oxford and New York: Berg.

Krippendorff, K. (1994) A recursive theory of communication. In D. Crowley and D. Mitchell (eds) *Communication Theory Today* (pp. 78–104). Cambridge: Polity Press.

Kristeva, J. (1991) *Strangers to Ourselves.* New York: Columbia University Press.

Lévi-Strauss, C. (1962) *La pensée sauvage.* Paris: Plon.

Lury, C. (1996) *Consumer Culture.* Cambridge: Polity Press.

Lury, C. (1997) The objects of travel. In C. Rojek and J. Urry (eds) *Touring Cultures: Transformations of Travel and Theory* (pp. 75–95). London and New York: Routledge.

Lyon, P. and Colquhoun, A. (1999) Selectively living in the past: Nostalgia and lifestyle. *Journal of Consumer Studies and Home Economics* 23, 191–196.

Lyon, P., Colquhoun, A., Kinney, D. and Murphy, P. (2000) Time travel: Escape from the late 20th century. *Jaargang* 18, 13–24.

MacCannell, D. (1976) *The Tourist: A New Theory of the Leisure Class.* New York: Schocken Books.

Macdonald, S. (1997) *Reimagining Culture: Histories, Identities and the Gaelic Renaissance.* Oxford and New York: Berg.

Malinowski, B. (1922) *Argonauts of the Western Pacific.* New York: Dutton.

Marcus, G.E. (1998) *Ethnography Through Thick and Thin.* Princeton, NJ: Princeton University Press.

Marx, K. (1859/1971) *A Contribution to the Critique of Political Economy* (S.W. Ryazanskaya, trans.). London: Lawrence & Wishart.

Mauss, M. (1990) *The Gift: The Form and Reason for Exchange in Archaic Societies.* New York and London: Routledge.

May, T. (1998) Reflections and reflexivity. In T. May and M. Williams (eds) *Knowing the Social World* (pp. 157–177). Buckingham: Open University Press.

May, T. and Williams, M. (1998) *Knowing the Social World.* Buckingham: Open University Press.

Mcfague, S. (1975) *Speaking in Parables: A Study in Metaphor and Theology*. London: SCM Press.

McGregor, A. (2000) Dynamic texts and tourist gaze: Death, bones and buffalo. *Annals of Tourism Research* 27 (1), 27–50.

Meethan, K. (2001) *Tourism in Global Society: Place, Consumption, Culture*. Basingstoke: Palgrave.

Merleau-Ponty, M. (2002) *Phenomenology of Perception*. London and New York: Routledge.

Miller, D. (1999) *Modernity, an Ethnographic Approach: Dualism and Mass Consumption in Trinidad*. Oxford: Berg.

Miller, D. (2000) Introduction: The birth of value. In P. Jackson, M. Lowe, D. Miller and F. Mort (eds) *Commercial Cultures* (pp. 77–83). Oxford: Berg.

Okri, B. (1996) *Birds Of Heaven*. London: Phoenix House.

Okri, B. (1997) *A Way of Being Free*. London: Phoenix House.

Olsen, K. (2002) Authenticity as a concept in tourism research: The social organization of the experience of authenticity. *Tourist Studies* 2 (2), 159–182.

Parker, M. (2002) *Against Management*. Oxford: Blackwell.

Parry, J. and Bloch, M. (eds) (1989) *Money and the Morality of Exchange*. Cambridge: Cambridge University Press.

Pi-Sunyer, O. (1978) Through native eyes. Tourists and tourism in a Catalan maritime community. In V. Smith (ed.) *Hosts and Guests: The Anthropology of Tourism* (pp. 148–155). Oxford: Blackwell.

Read, A. (1993) *Theatre and Everyday Life: An Ethics of Performance*. London: Routledge.

Reid, I. (1992) *Narrative Exchanges*. London and New York: Routledge.

Richards, G. and Wilson, J. (2004) *The Global Nomad*. Clevedon: Channel View.

Ricoeur, P. (1984) *Time and Narrative*. Chicago, IL: University of Chicago Press.

Ritzer, G. (1996) *The McDonaldization of Society: An Investigation into the Changing Character of Contemporary Social Life* (revised edn). Thousand Oaks, CA: Forge Pine Press/Sage.

Ritzer, G. (1997) *McDonaldization: Explorations and Extensions*. London: Sage.

Ritzer, G. and Liska, A. (1997) 'McDisneyization' and 'post-tourism': Complementary perspectives on contemporary tourism. In C. Rojek and J. Urry (eds) *Touring Cultures: Transformations of Travel and Theory* (pp. 96–112). London and New York: Routledge.

Robinson, M. and Andersen, H.C. (2002) *Literature and Tourism*. London: Continuum.

Rojek, C. (1993) *Ways of Escape*. London: Macmillan.

Rose, D. (1990) *Living the Ethnographic Life*. London: Sage.

Sahlins, M. (1972) *Stone Age Economics*. Chicago, IL: Aldine.

Schechner, R. (1982) Collective reflexivity: Restoration of behavior. In J. Ruby (ed.) *A Crack in the Mirror: Reflexive Perspectives in Anthropology* (pp. 39–81). Philadelphia, PA: University of Pennsylvania Press.

Schiffer, M.B. and Miller, A.R. (1999) *The Material Life of Human Beings: Artifacts, Behavior and Communication*. London: Routledge.

Scott, J. (1990) *Domination and the Arts of Resistance: Hidden Transcripts*. New Haven, CT: Yale University Press.

Simms, K. (2003) *Paul Ricoeur*. London: Routledge.

Smith, V. (ed.) (1997) *Hosts and Guests*. Oxford: Blackwell.

Stagl, J. (1980) Die Apodemik oder 'Reisekunst' als Methodik der Sozialforschung vom Humanismus bis zur Aufklärung. In M.J. Rassem and J. Stagl (eds) *Statistik und Staatbeschreibung in der Neuzeit* (pp. 131–202). Paderborn: F. Schöningh.

Turner, V. (1982) *From Ritual to Theatre: The Human Seriousness of Play*. New York: PAJ Publications.

Turner, V. and Turner, E. (1978) *Image and Pilgrimage in Christian Culture; Anthropological Perspectives*. Thousand Oaks, CA: Pine Forge Press/Sage.

Urry, J. (1990) *The Tourist Gaze: Leisure and Travel in Contemporary Societies*. London: Sage.

Venuti, L. (1995) *The Translator's Invisibility: A History of Translation*. London and New York: Routledge.

Wang, N. (1999) Rethinking authenticity in tourism experience. *Annals of Tourism Research* 26 (2), 349–370.

Wang, N. (2000) *Tourism and Modernity: A Sociological Analysis*. Oxford: Pergamon.

Weber, M. (1978) *Economy and Society*. New Jersey: Bedminster Press.

Wierlacher, A. (ed.) (1993) *Kulturthema Fremdheit: Leitbegriffe und Problemfelder kulturwissenschaftlicher Fremdheitsforschung*. Munich: Iudicium.

Wilk, R. (1995) Learning to be local in Belize: Global systems of common difference. In D. Miller (ed.) *Worlds Apart: Modernity through the Prism of the Local* (pp. 110–133). London: Routledge.

Williams, R. (1977) *Marxism and Literature*. Oxford: Oxford University Press.

Williams, R. (2000) *Lost Icons: Reflections on Cultural Bereavement*. London and New York: T&T Clark.

Index